Tissue Repair and Reconstruction

Series Editors

Andy H. Choi, Carlingford, NSW, Australia

Besim Ben-Nissan, Sydney, NSW, Australia

SpringerBriefs in Tissue Repair and Reconstruction provides a unique perspective and in-depth insights into the latest advances and innovations contributing to improved and better treatments for patients with damaged soft and hard tissues as a result of diseases, trauma, and implantations. The book series consists of volumes that offer biomedical researchers better insights into the advancements of biomaterials science and their translation from the laboratory to a clinical setting. Similarly, the series provides information to surgeons and medical practitioners on novel ideas in biomedical science and engineering on top of disseminating new ideas and know-hows in diagnostics and treatment options for patients from head to toe.

The series will cover a number of key topics:

Fundamental Concepts and Surface Modifications: The topic will provide detailed information on the discovery and advancements of biomaterials surface modification approaches and their use within the human body in a safe manner and without provoking any negative tissue response.

Computational Simulations and Biomechanics: Anatomically accurate computational models are being in all fields of medicine particularly in orthopedics and dentistry to reveal the biomechanical functions and behaviors of bones and joints when damaged, diseased, and in the health state. They also contribute to our understanding during the design and applications of implants and prosthetics subjected to functional loadings and movements.

Surgical Advances and Treatment Options: Discusses how surgical techniques are revolutionized by our deeper understanding into biomaterials science and tissue engineering. The section also focuses on the latest innovations and surgical advancements currently being used to treat patients with damaged tissues.

Post-Operative Treatment and Rehabilitation Engineering: Expands the independence and functionality of the patient after surgery while at the same time reducing the chance of complications such as wound infections and dislocations. Advances in technologies are creating new opportunities in how physiotherapy rehabilitations are delivered.

Andy H. Choi · Sian Yik Lim
Editors

Pharmacological Interventions for Osteoporosis

 Springer

Editors
Andy H. Choi
School of Life Sciences
Translational Biomaterials and Medicine
Group
University of Technology
Sydney, NSW, Australia

Sian Yik Lim
Hawaii Pacific Health
Straub Clinic
Honolulu, HI, USA

ISSN 2731-9180   ISSN 2731-9199 (electronic)
Tissue Repair and Reconstruction
ISBN 978-981-99-5825-2   ISBN 978-981-99-5826-9 (eBook)
https://doi.org/10.1007/978-981-99-5826-9

This Springer imprint is published by the registered company Springer Nature Singapore Pte Ltd.
The registered company address is: 152 Beach Road, #21-01/04 Gateway East, Singapore 189721,
Singapore

Paper in this product is recyclable.

Preface

Preventing osteoporosis and its associated fractures is regarded as crucial in maintaining the independence, health, and quality of life of the elderly because of the maleficent effects of osteoporosis. Age-related bone loss is asymptomatic, and the morbidity of osteoporosis is secondary to the fractures that take place. Osteoporosis results in skeletal fragility along with a heightened risk of fracture caused by a change in bone remodeling. It has been accepted that the principle behind the bone remodeling sequence is to sustain the integrity of the skeleton. This is accomplished through the collaborative efforts of osteoclasts (responsible for bone resorption) and osteoblasts (responsible for bone formation). Bone resorption and bone formation are balanced in homeostatic equilibrium. On a cellular level, simultaneous mechanical and biological actions play a governing role in the delicate equilibrium between bone growth, formation, and resorption. In osteoporosis, however, the quantity of resorbed bone by osteoclasts exceeds the amount of bone formed by osteoblasts resulting in a reduction in bone strength and damage to the skeletal architecture.

The purposes of intervention are to avoid bone loss in patients diagnosed with osteoporosis as well as in individuals at risk of osteoporosis. Treatments may be targeted at maintaining bone mass or repairing skeletal deficits. The objectives of interventions and treatment focus are identical, that is to decrease the chance of osteoporotic fractures by maximizing skeletal strength. Changes in lifestyle are helpful but patients that have a high risk of fracture will often also require pharmacological interventions. A number of biological studies have revealed the mechanisms governing bone remodeling and consequently resulted in the discovery of new pharmacological targets that could aid in enhancing bone health in patients diagnosed with osteoporosis. The medications used in the treatment of osteoporosis can be categorized into either primarily anabolic or anti-resorptive. Anabolic pharmaceutical agents are responsible for promoting new bone formation, whereas anti-resorptive pharmaceutical agents are responsible for the prevention of bone resorption. Due to their capacity to selectively restrain the activity of osteoclast and ultimately slow down bone resorption, bisphosphonate therapy has become the primary clinical intervention for postmenopausal osteoporosis for the past two decades. On the other hand, there are potential concerns and complications associated with their widespread use in the clinical environment such as osteonecrosis of the jaw. Recently, two monoclonal antibodies romosozumab and denosumab were introduced based on the discovery of sclerostin in restricting the differentiation of osteoblasts and the formation of osteoclast sustained by the receptor activator of nuclear factor kappa-B ligand (RANKL). In the United States, teriparatide and abaloparatide are the two other anabolic pharmaceutical agents approved to treat osteoporosis and they have been shown to reduce the incidence of non-vertebral and vertebral fractures significantly after they were administered to patients daily as subcutaneous injections.

Written by international experts from different specialties based in Australia, Italy, Malaysia, and the United States, it is envisaged that this book will provide readers with fundamental insights into the basic properties of the pharmaceutical agents used in the treatment of osteoporosis as well as their mechanisms of action and clinical outcomes. We also include topics covering the clinical application of osteoporosis treatments that would be of interest to a wide range of audiences involved in osteoporosis care including primary care physicians, endocrinologists, rheumatologists, orthopedic surgeons, and dentists.

Finally, we would like to express our deepest gratitude to all our contributing authors and the people at Springer Publishing, especially Dr. Ramesh Premnath, Ramamoorthy Rajangam, and Mano Priya Saravanan for their help and for making the book possible.

Sydney, Australia   Andy H. Choi
Hawaii, USA   Sian Yik Lim

Contents

About the Editors

Dr. Andy H. Choi is an early career researcher who received his Ph.D. from the University of Technology Sydney (UTS) in Australia in 2004 on the use of computer modelling and simulation known as finite element analysis (FEA) to examine the biomechanical behavior of implants installed into a human mandible. After completing his Ph.D., he expanded his research focus from FEA to sol-gel synthesis of multifunctional calcium phosphate nano coatings and nano composite coatings for dental and biomedical applications.

In late 2010, Dr. Choi was successfully awarded the internationally competitive Endeavour Australia Cheung Kong Research Fellowship Award and undertook post-doctoral training at the Faculty of Dentistry of the University of Hong Kong focusing on the application of FEA in dentistry and the development of calcium phosphate nano-bioceramics.

He is currently serving as an associate editor for the Journal of the Australian Ceramic Society and as an editor for a number of dentistry-related journals. In addition, he is also serving as an editorial board member for several dentistry, nanotechnology, and orthopedics journals. To date, Dr. Choi has published over 50 publications including 5 books and 30 book chapters on calcium phosphate, nano-biomaterial coatings, sol-gel technology, marine structures, drug delivery, tissue engineering, and finite element analysis in nanomedicine and dentistry.

Dr. Sian Yik Lim is a rheumatologist at Pali Momi Bone and Joint Center, Hawaii Pacific Health. He currently runs a specialty osteoporosis clinic at Pali Momi Bone and Joint Center and has been involved in efforts to improve the quality of osteoporosis care in Hawaii. He is a clinical densitometrist certified by the International Society for Clinical Densitometry. After graduating from Osaka University School of Medicine, he subsequently trained in internal medicine and rheumatology in the United States. He completed his rheumatology fellowship at Massachusetts General Hospital and was a research fellow at Harvard Medical School. He has received the American Federation of Medical Research Resident Research Day Award and

Marshal J Schiff, MD Memorial Fellow Research Award for his research. He has published more than 30 papers and abstracts about gout, osteoporosis, and septic arthritis in respected journals such as JAMA, rheumatology, and current opinions in rheumatology.

Bisphosphonates: Clinical Applications and Perspectives in Osteoporosis Treatment

Sian Yik Lim and Marcy B. Bolster

Abstract Osteoporosis is a skeletal disorder characterized by decreased bone strength, leading to increased fracture risk. In this article, we discuss the use of bisphosphonates in the treatment of osteoporosis. We aim to give the reader a strong background of using bisphosphonates in osteoporosis treatment. We also discuss important topics pertinent to clinical care, including the potential side effects and adverse effects, as well as strategies to mitigate them. We also describe the long-term use of bisphosphonates with efficacy and safety in mind. We hope to provide clinicians with information that will be useful in daily practice when prescribing bisphosphonates for the treatment of osteoporosis.

Keywords Osteoporosis · Bisphosphonates · Skeletal disorder

1 Introduction

Osteoporosis is a skeletal disorder, characterized by reduced bone strength, leading to an increased fracture risk [1]. Fragility fractures related to osteoporosis are associated with significant morbidity, mortality, and health care costs. The management of osteoporosis focuses on reducing fracture risk. This article discusses bisphosphonates, one of the first medicines used to treat osteoporosis.

S. Y. Lim (✉)
Bone and Joint Center, Pali Momi Medical Center, 98-1079 Moanalua Road, Suite 300, Aiea, HI 96701, USA
e-mail: limsianyik@gmail.com

Hawaii Pacific Health Medical Group, Honolulu, HI, USA

Department of Family Medicine, John A Burns School of Medicine, University of Hawaii, Honolulu, HI, USA

M. B. Bolster
Rheumatology Fellowship Training Program, Massachusetts General Hospital, Boston, MA, USA

Harvard Medical School, Boston, MA, USA

While bisphosphonates were first synthesized in the 1800s, their use in medicine started in the 1960s. In the 1960s, William Neuman and Herbert Fleisch identified inorganic pyrophosphate in the urine and serum of study subjects. They postulated that inorganic pyrophosphate was potentially a natural water softener that prevented soft tissue calcification and potentially could be used to treat osteoporosis. However, pyrophosphate was only active when injected and not active orally because of the hydrolysis of pyrophosphate in the gastrointestinal tract.

Bisphosphonates, on the other hand, are stable even when given orally. This stability was a key reason for bisphosphonates being developed for medical uses. Bisphosphonates, like inorganic pyrophosphate, not only inhibited calcification in the human body but also had unique properties of inhibiting calcium phosphate dissolution [2]. This property was extrapolated for the possible treatment of bone disease, with studies evaluating their potential to inhibit bone resorption.

2 Chemical Structure of Bisphosphonates

2.1 Basic Chemical Structure of Bisphosphonates

Bisphosphonates are chemically stable derivatives of inorganic pyrophosphate, and the general structure is shown in Fig. 1 (left). Inorganic pyrophosphate (Fig. 1 (right)) is a naturally occurring compound in the body, released as a byproduct of synthetic reactions in the body [3]. Pyrophosphate inhibits calcification by binding to hydroxyapatite crystals in bone. However, because it is degraded rapidly by pyrophosphatases, it exhibits very little biological activity in vivo. Substitution of the oxygen atom in pyrophosphate by a carbon atom (Fig. 1a) produces a chemically stable bisphosphonate structure.

Two carbon phosphorus (C-P) bonds sharing a single carbon atom (P-C-P) are called germinal bisphosphonates resistant to enzymatic hydrolysis [3, 4]. Most bisphosphonates used in clinical practice have a hydroxyl group in the R1 position. The phosphate and hydroxyl groups are essential for the affinity of bisphosphonates

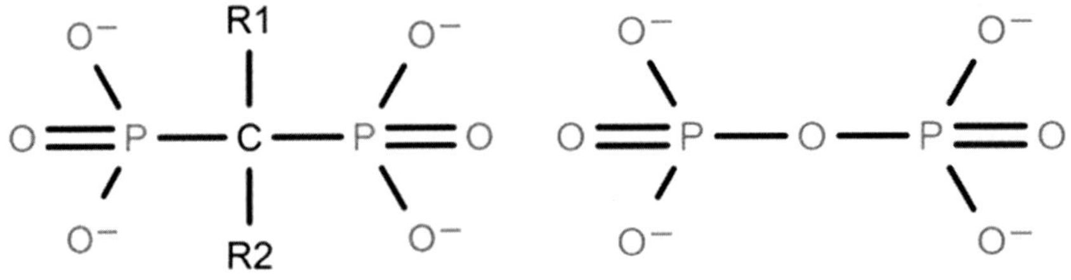

Fig. 1 The general chemical structure of bisphosphonate (left), and inorganic pyrophosphate (right)

for the bone matrix. The phosphate groups have a high affinity towards bone hydrox-yapatite crystals, while the hydroxyl group enhances the ability of bisphosphonates to bind to calcium [3].

2.2 Chemical Structures of Different Bisphosphonates Used in Clinical Practice

Modifying the moiety structure bound to R2 enabled the development of different bisphosphonates. The structural moiety attached to R2 determines the potency of bisphosphonates in inhibiting bone resorption. The first-generation bisphosphonates were the non-nitrogen bisphosphonates (i.e., clodronate, etidronate), with the structural moiety in R2 with no nitrogen or amino groups attached. Subsequently, introducing nitrogen or amino groups in the R2 structural moiety led to the development of potent bisphosphonates (i.e., alendronate, risedronate, ibandronate and zoledronic acid) with higher antiresorptive activity, many of which are currently used in the treatment of osteoporosis [5].

3 Important Pharmacokinetic Properties of Bisphosphonates for Clinical Practice

Bisphosphonates have low bioavailability, especially when administered orally. Minimal quantities of the oral dose (1–4%) are absorbed through the gastrointestinal tract [6]. The poor gastrointestinal absorption relates to bisphosphonates being poorly lipophilic. After absorption within the body, approximately 50% of the absorbed dose binds to hydroxyapatite within the bone and coats the bone surface [3]. The other 50% of absorbed bisphosphonates, not attached to bone, are excreted rapidly in the urine [4]. Factors that affect bisphosphonate skeletal retention include patient factors such as comorbidities (i.e., kidney function), the baseline rate of bone turnover, and drug-related factors (i.e., the potency of bisphosphonate). Bisphosphonates have a very high affinity for bone tissue; other tissues in the body do not take up bisphosphonates, and bisphosphonates are not metabolized by the body [7]. Bisphosphonates have a long half-life, ranging from 1 to 10 years, and the half-life largely depends on the bone turnover rate [5].

Oral bisphosphonates are chelating agents that form salts with multivalent cations in food, dietary supplements, and medications. The formation of insoluble salts prevents gastrointestinal absorption. Therefore, it is recommended to have a 30-min to a 2-h interval between oral bisphosphonate administration and ingestion of vitamins, dietary supplements, and antacids that contain multivalent cations (i.e., calcium, magnesium) [8].

4   Mechanism of Action of Bisphosphonates

4.1   Inhibition of Osteoclasts

The uptake of bisphosphonates within the skeleton is not homogeneous. Bisphosphonates are preferentially incorporated into the bone where there is high turnover [5]. Osteoclasts internalize bisphosphonates on the bone surface. Non-nitrogen-containing bisphosphonates are incorporated into newly formed adenosine triphosphate (ATP) within osteoclasts. These ATP analogs are non-hydrolyzable and are cytotoxic to osteoclasts. Accumulation of these ATP analogs leads to apoptosis of the osteoclasts. Nitrogen-containing bisphosphonates inhibit osteoclasts by a different mechanism. These bisphosphonates block farnesyl pyrophosphate (FPP) synthase in the mevalonate pathway. This leads to decreased levels of farnesyl diphosphate and geranylgeranyl diphosphate, which play a role in the prenylation of small GTPase proteins critical for osteoclast survival and function [9]. Furthermore, by blocking FPP synthase, nitrogen-containing bisphosphonates increase the formation of an ATP analog 1-adenosine-5′-yl ester 3-(-3methylbut-3-enyl) ester triphosphoric acid (ApppI). ApppI induces osteoclast apoptosis and does not have a bisphosphonate within its structure [9].

4.2   Bisphosphonates as Anti-remodeling Agents

Bone resorption and formation are coupled in bone remodeling of the normal skeleton. Bisphosphonates, by inhibiting osteoclasts, indirectly inhibit osteoblasts. Therefore, bone formation is also decreased (less than bone resorption) when bisphosphonates inhibit bone resorption. This effect can be noted from the change in bone turnover markers after administration of intravenous zoledronic acid, for example, where there is a rapid decrease of bone resorption markers associated with a smaller gradual reduction in bone formation markers [10]. Inhibition of bone resorption and formation is also noted with other anti-remodeling agents, such as denosumab. With the parathyroid hormone analogs, there is stimulation of both bone formation and bone resorption. Uniquely, with romosozumab, there is increased bone formation with decreased bone resorption due to the decoupling of bone remodeling [11].

5   Efficacy

Several clinical trials established the efficacy of bisphosphonates [10, 12–18]. Initial trials were limited to postmenopausal women, but subsequent studies confirmed anti-fracture efficacy in men and in glucocorticoid induced osteoporosis [4]. In general, bisphosphonates show greater effectiveness in preventing vertebral fractures as compared to non-vertebral fractures. As previously described, bisphosphonates are preferentially incorporated into bone with high turnover, and osteoclasts internalize bisphosphonates on the bone surface. Vertebral sites have more trabecular bone where bone turnover is more active.

Oral bisphosphonates are considered first-line treatment for osteoporosis due to fracture-risk reduction efficacy and to their cost-effectiveness. Of the three oral bisphosphonates, because alendronate and risedronate reduce the risk of vertebral fractures, hip fractures, and non-vertebral fractures (see Sects. 5.1 and 5.2), they are favored as first-line oral treatment by many practitioners, especially in patients at high risk for fracture. Ibandronate has not been shown in randomized control trials to reduce non-vertebral fracture risk (see Sect. 5.3).

Due to convenience, intermittent dosing regimens, such as weekly alendronate or risedronate, are reasonable choices as initial therapy. The efficacy of intermittent dosing regimens is equivalent to daily dosing regimens (see Sect. 5.5). For patients favoring once-monthly treatment, oral risedronate 150 mg monthly may be a better option than oral ibandronate 150 mg monthly [19], because of more robust data supporting antifracture efficacy in vertebral and non-vertebral fractures.

5.1 Alendronate

Alendronate is effective in reducing vertebral fracture risk in postmenopausal women. Alendronate also effectively reduces the risk of non-vertebral fragility fractures and hip fractures.

The efficacy of alendronate was demonstrated in the pivotal Fracture Intervention Trial (FIT) trial [12]. FIT was a randomized, double-blind, placebo-controlled, parallel-group study to evaluate the efficacy of alendronate in reducing the risk of vertebral fractures (in osteoporosis trials, vertebral fractures are diagnosed based on vertebral morphometry: vertebral fractures diagnosed based on semiquantitative assessment or quantitative measurement of vertebral dimensions) and clinical fractures in women with low bone mass.

The FIT trial included postmenopausal women aged 55–81 years with low femoral neck bone mineral density (BMD). The study participants had a femoral neck BMD of 0.68 g/cm^2 or less. The study included a placebo group with two study groups: one group with a history of existing vertebral fracture (The Vertebral Deformity Study) [12] and one group with no prior vertebral fracture (The Clinical Fracture Study) [13]. The primary endpoint of the Vertebral Deformity Study was the incidence of new vertebral fractures at three years. The primary endpoint of the Clinical Fracture Study was the incidence of all clinical fractures during a follow-up period of 4–4.5 years.

In the Vertebral Deformity Study, 2027 subjects were randomized into two groups with 1022 receiving alendronate, and 1005 subjects receiving placebo. The dose of alendronate was 5 mg daily initially, and then increased to 10 mg daily at 24 months. The average age of patients in the study was approximately 71 years, with well-matched baseline characteristics between both treatment arms. The primary endpoint of morphometric vertebral fracture risk reduction was achieved. At the end of the study period, alendronate led to a 47% reduction in new vertebral fractures. Alendronate showed a 38% relative rate reduction (RRR) of clinical fractures, 51% RRR of hip fractures, and 48% RRR of forearm fractures compared to the placebo group; compared to placebo, alendronate treatment increased BMD by 6.2% at the lumbar spine and 4.7% at the total hip. The Clinical Fracture Study is summarized in Table 1, although alendronate treatment increased BMD at all sites, the primary endpoint was not met: alendronate treatment did not significantly reduce the rate of all clinical fractures at 4.5 years (14% risk reduction).

5.2 Risedronate

Risedronate increases BMD and it reduces vertebral and non-vertebral fragility fractures in postmenopausal women with osteoporosis. Vertebral Efficacy with Risedronate Therapy-North America (VERT-NA) and Vertebral Efficacy with Risedronate Therapy-North America-Multi-National (VERT-MN) trials demonstrated the efficacy of risedronate in reducing vertebral fractures [14, 15] (Table 1). Risedronate reduced hip fracture risk in the Hip Intervention Program Trial [16] (Table 1) in patients with confirmed osteoporosis. Risedronate did not reduce fractures in older women selected based on risk factors other than low bone mineral density [16].

5.3 Ibandronate

Ibandronate has been demonstrated to reduce vertebral fragility fracture rates in postmenopausal women with osteoporosis. While a reduction in vertebral fragility fractures was noted in the pivotal Oral Ibandronate Osteoporosis Vertebral Fracture Trial in North America and Europe (BONE) trial, there was no difference in non-vertebral fragility fractures in patients who received both dosing regimens of 2.5 mg daily and 20 mg every other day for 12 doses every three months [17]. In post hoc analyses of BONE, a 69% reduction in non-vertebral fracture risk was noted in higher-risk patients with a T-score of less than −3.0. Due to less robust data for ibandronate reducing the risk of non-vertebral fractures, ibandronate may be a less favorable choice in patients at higher risk of non-vertebral fractures.

5.4 Zoledronic Acid

Zoledronic acid has been shown to reduce vertebral fractures, non-vertebral fractures, and hip fractures, as well as demonstrating a gain in BMD. The efficacy of zoledronic

Table 1 Efficacy studies of bisphosphonates (BMD: bone mineral density, LS: lumbar spine, FN: femoral neck, TN: total neck)

Study name	Trial design/study context	Study population	Comparator groups/ denosumab dose	Primary endpoint	Conclusions
Alendronate					
Fracture intervention trial (FIT)-1 [12]	Randomized control trial	2027 postmenopausal women with ≥1 vertebral fracture	Oral alendronate for 3 years or placebo **Alendronate doses**: 5 mg daily for 2 years, 10 mg daily for the remainder of the trial	(1) New vertebral fracture (X-ray) at 3 years	(1) New vertebral fracture reduced by 47% (2) Hip fracture reduced by 51% (3) BMD gain compared to placebo: LS: 6.2%, FN 4.1%, TH 4.7%
Fracture intervention trial-2 [13]	Randomized control trial	4432 postmenopausal women with low BMD ≤0.68 g/cm² with no vertebral fractures at baseline	Oral alendronate for 4 years or placebo **Alendronate doses**: 5 mg daily for 2 years, 10 mg daily for the remainder of the trial	(1) New Clinical fractures at 4 years	(1) Reduction in all clinical fractures not statistically significant (2) New vertebral fracture reduced by 44%

(continued)

Table 1 (continued)

Study name	Trial design/study context	Study population	Comparator groups/denosumab dose	Primary endpoint	Conclusions
Risedronate					
Vertebral efficacy with risedronate-North America (VERT-NA) [14]	Randomized control trial **Study Context**: Pivotal trial for risedronate: fracture prevention efficacy and safety study	2458 postmenopausal women with at least 1 vertebral fracture at baseline in North America	Oral risedronate for three years or placebo **Risedronate doses**: 2.5 mg, 5 mg daily *2.5 mg dose arm discontinued after 1 year due to 2.5 mg daily less effective than 5 mg daily in other trials	(1) New vertebral fracture (X-ray) at 3 years	(1) New vertebral fracture reduced by 41% (2) Non-vertebral fracture reduced by 39% Mean BMD change from baseline treatment: LS: 5.4%, TH: 3.3%, FN: 1.6% Mean BMD change from baseline, placebo LS: 1.1%, TH: −0.7%, FN: −1.2%
Vertebral efficacy with risedronate-North America (VERT-MN) [15]	Randomized control trial **Study context**: Pivotal trial for risedronate: fracture prevention efficacy and safety study	1126 postmenopausal women with osteoporosis with 2 or more prevalent fractures in Europe and Australia	Oral Risedronate for three years or placebo **Risedronate doses**: 2.5 mg then 5 mg daily *2.5 mg dose arm discontinued after 1 year due to 2.5 mg daily less effective than 5 mg daily in other trials	(1) New vertebral fracture (X-ray) at 3 years	(1) New vertebral fracture reduced by 49% (2) Non-vertebral fracture reduced by 33% (3) BMD gain compared to placebo: LS: 5.9%, TH 6.4%

(continued)

Table 1 (continued)

Study name	Trial design/study context	Study population	Comparator groups/denosumab dose	Primary endpoint	Conclusions
Hip intervention trial [16]	Randomized control trial	5545 women age 70–79 years with osteoporosis (FN T-score < −4, < −3 with non-skeletal risk factor for hip fracture (poor gait, or likely to fall), 3886 women age 80 or older, with clinical risk factors but no low BMD	Oral risedronate for three years or placebo **Risedronate doses:** 2.5 mg or 5 mg daily	(1) Hip fracture at 3 years	(1) Overall hip fracture rate reduced by 30% ($p = 0.02$) (2) Patients with osteoporosis, hip fracture reduced by 40% ($p = 0.009$) (3) Patients with clinical risk factors, hip fracture reduced by 4.2% ($p = 0.35$)
Ibandronate					
Oral ibandronate osteoporosis vertebral fracture trial in North America and Europe (BONE) [17]	Randomized control trial	2946 postmenopausal women age 55–80 years, 1–4 prevalent vertebral fractures and T-score −2 to −5 in more than 1 vertebral bodies	Oral ibandronate for three years or placebo **Risedronate doses:** 2.5 mg daily (daily dosing group) 20 mg every other day for 12 doses every 3 months (intermittent dosing group)	(1) New vertebral fracture (X-ray) at 3 years	(1) New vertebral fracture reduced by 49% (2) Non-vertebral fracture reduced by 33% BMD changes (3) BMD gain compared to placebo: LS: 5.9%, TH 6.4%

(continued)

Table 1 (continued)

Study name	Trial design/study context	Study population	Comparator groups/ denosumab dose	Primary endpoint	Conclusions
Zoledronic acid					
Health outcomes and reduced incidence with zoledronic acid once yearly (HORIZON) pivotal fracture trial [10]	Randomized control trial	7736 postmenopausal women, femoral neck T-score −2.5 or less, or femoral neck T-score −1.5 or less with at least 2 mild vertebral fractures or one moderate vertebral fracture Stratum 1-concomitant use of osteoporosis medications at baseline (hormone therapy, raloxifene, calcitonin, tibolone, tamoxifen, dehydroepiandrosterone, ipriflavone, and medroxyprogesterone) Stratum 2-no concomitant use of osteoporosis medications	Intravenous zoledronic acid or placebo **Zoledronic acid doses:** 5 mg annually, for 3 years	(1) New vertebral fracture (X-ray) at 3 years (Stratum 2) (2) New hip fracture incidence at 3 years (Stratum 1)	(1) New vertebral fracture reduced by 70% in Stratum 1 patients, $p < 0.001$ (2) Hip fracture reduced by 41% in all patients. $P = 0.002$ (3) BMD gain compared to placebo: LS: 6.7%, FN 5.1%, TH 6.0%

(continued)

Table 1 (continued)

Study name	Trial design/study context	Study population	Comparator groups/denosumab dose	Primary endpoint	Conclusions
Health outcomes and reduced incidence with zoledronic acid once yearly (HORIZON) recurrent fracture trial	Randomized control trial	2664 patients with recent hip fracture (men and women)	Intravenous zoledronic acid or placebo **Zoledronic acid doses:** Intravenous 5 mg single dose	(1) New clinical fracture, excluding facial and digital fractures or fractures in abnormal bone Average follow-up time 1.9 years	(1) New clinical fracture reduced by 35% (2) Mortality reduced by 28%

acid was demonstrated in the HORIZON Pivotal Fracture trial. HORIZON was a randomized, double-blind, placebo-controlled study to assess the fracture reduction efficacy of zoledronic acid 5 mg administered annually [10].

The HORIZON trial included 7736 postmenopausal women with osteoporosis. Inclusion criteria required either a femoral neck T-score <-2.5 or a femoral neck T-score <-1.5 with at least two mild vertebral fractures or one moderate vertebral fracture. In this study, 3833 patients were randomized to receive yearly IV zoledronic acid infusions, and 3876 patients were randomized to receive placebo.

The study participants were divided into two strata: Stratum 1-concomitant use of osteoporosis medications at baseline and Stratum 2-no concomitant use of osteoporosis medications. The primary endpoint was the incidence of new vertebral deformity (patients in stratum 2) and hip fracture incidence (all patients) at three years.

At year 3, morphometric vertebral fractures were reduced by 70% in Stratum 1 patients. Hip fracture in all patients was reduced by 41% at three years. Also noted at three years was a 77% reduction in clinical vertebral fractures and a 25% reduction in non-vertebral fractures. Zoledronic acid administration was associated with a gain in BMD at the lumbar spine, femoral neck, and total hip of 6.7%, 5.1%, and 6.0%, respectively, compared to the placebo group.

The HORIZON-Recurrent Fracture Trial assessed zoledronic acid efficacy in the secondary prevention of clinical fractures in patients who had sustained a hip fracture [18]. Patients received zoledronic acid intravenously within 90 days of surgical repair for hip fracture. Zoledronic acid was given to 1065 patients, and 1062 patients were assigned to placebo. The primary endpoint was the development of new clinical fractures. Clinical fractures were reduced by 35% in the treatment group. Zoledronic acid reduce mortality from any cause was reduced by 28% [20]. Although zoledronic acid reduces fracture rates and fractures are associated with increased mortality, only 8% of zoledronic acid effect on mortality was explained by the prevention of subsequent fractures. Mortality rate reductions related additionally to reduce cardiovascular events and pneumonia [20].

5.5 Bridging Studies for Intermittent Dosing Efficacy

While previously mentioned trials established the efficacy for daily dosing of alendronate, risedronate, and ibandronate, bridging studies have demonstrated equivalent effects on BMD for less frequent dosing. In bridging studies, non-inferiority trials of shorter duration have shown equal BMD gains and bone turnover marker changes with daily dosing. Bridging studies demonstrating similar efficacy led to the approval of alendronate 70 mg weekly [21], risedronate 35 mg weekly [22], and risedronate 150 mg monthly [23]. In the MOBILE study, ibandronate 150 mg monthly led to more significant gains in BMD than daily dosing [24]. In the Dosing Intravenous Administration trial, intravenous ibandronate 3 mg every three months also led to more significant gains in BMD compared to daily oral dosing of ibandronate [25]. Both doses of ibandronate were approved for the treatment of osteoporosis.

6 Adverse Events

6.1 Gastrointestinal Side Effects

Despite a few observational studies [26] and anecdotal impressions of some clinicians noting a higher rate of intolerance to bisphosphonates due to gastrointestinal irritation, no difference has been reported between subjects treated with bisphosphonates compared to those treated with placebo in randomized control trials [27]. As described in the Federal Drug Administration (FDA) label, bisphosphonates administered orally may cause local irritation of the upper gastrointestinal mucosa [28]. Due to the possibility of worsening underlying disease, bisphosphonates should be used with caution in patients with an underlying gastrointestinal disorder such as esophagitis, Barrett's esophagus, dysphagia, other esophageal diseases, gastritis, duodenitis, or ulcer disease [28]. Oral bisphosphonates may be used in patients with well-controlled gastroesophageal reflux or peptic ulcer disease [19]. These medications, however, are relatively contraindicated in patients with active esophageal disorders such as esophageal varices with a high risk for bleeding, esophageal stricture, and achalasia. Based on postmarketing surveillance reports of the possibility of erosive esophagitis when not taking the medication properly, it is recommended that patients swallow the pill with a full glass of water (about 240 mL), avoid lying down for at least 30 min, and wait 30 min to ingest their first food of the day [29]. Weekly dosing regimens may have fewer gastrointestinal side effects when compared to daily dosing [30]. While observational studies have shown that risedronate may have fewer gastrointestinal side effects than alendronate, placebo-controlled trials have not demonstrated a significant difference [30]. Nevertheless, prescribing risedronate for a patient who had gastrointestinal side effects with alendronate can be considered [19].

6.2 Acute Phase Reactions

Acute phase reactions are transient inflammatory states characterized by influenza-like symptoms such as fever, myalgia, and arthralgia [31]. Acute phase reactions occur in approximately 40% of patients after intravenous zoledronic acid infusion and also occur, though less commonly with oral bisphosphonates (about 5%) [31]. Before starting bisphosphonate therapy, it is important to advise patients about this possibility. Prophylactic treatment with acetaminophen or nonsteroidal anti-inflammatory drugs (NSAIDs) can be considered, especially in patients receiving intravenous zoledronic acid; prophylactic treatment with these agents has been shown to reduce acute phase reaction symptoms by 30–40% [31]. One could consider advising patients to take acetaminophen 650–1000 mg 30–60 min prior to the infusion. Patients with a history of acute phase reactions with intravenous zoledronic acid can be advised to take tylenol 650 mg 30–60 min before infusion, then every 6–8 h for 3 days [31]. Patients should be advised that acute phase reactions decrease in frequency with subsequent infusions.

6.3 Medication-Related Osteonecrosis of the Jaw (MRONJ)

Osteonecrosis of the jaw (ONJ) is defined as an area of exposed bone in the maxillofacial area where there is no healing over eight weeks after being recognized by a health care provider in the setting of exposure to an antiresorptive agent [32]. Risk factors for ONJ include infections, and trauma or injury to the jawbones as associated with dentoalveolar surgery [33]. The incidence of ONJ in patients with osteoporosis is low, ranging from 0.15 to 0.001% person-years of exposure, and is only slightly higher than that occuring in the general population. The incidence of ONJ is higher in the oncology population where antiresorptive agents are more commonly used at higher doses. Before initiation of bisphosphonates, inquiries about upcoming invasive dental procedures should be performed. Consideration should be given to delaying bisphosphonate treatment for such patients to provide a window of healing [19], and adequacy of healing should be ensured after the procedure. Other recommended precautions for patients include emphasizing the importance of maintaining good oral hygiene and consideration for antibiotics/antimicrobial rinses to use pre- and post-dental procedures, especially in high-risk patients [32]. Regarding patients who are already taking bisphosphonate therapy and develop the need for invasive procedures, there is no significant data to guide clinical decisions of whether to stop or continue treatment. The American Association of Oral and Maxillofacial Surgeons Position Paper in 2014 has suggested stopping bisphosphonate therapy in high-risk patients (cumulative bisphosphonate exposure of more than four years, patients with rheumatoid arthritis, diabetes, tobacco use, and/or glucocorticoid exposure) approximately two months before undergoing an invasive dental procedure and until the site is healed [34]. Management of mild-moderate cases is usually conservative, including good dental hygiene, antimicrobial rinses, and systemic antibiotic treatment, as indicated [32].

6.4 Atypical Femoral Fractures

Atypical femoral fractures, characterized by fractures in the subtrochanteric region and diaphysis of the femur, have unique features that differ from femoral diaphyseal osteoporotic fractures [35, 36]. These features include (1) no associated trauma/ minor trauma, (2) fracture line originating from the lateral cortex and occurring in a transverse plane, (3) complete fracture extends through both cortices and may be associated with a medial spike, (4) minimally/non-comminuted, and (5) lateral periosteal or endosteal thickening of the lateral cortex present at the fracture site. Although the precise etiopathogenesis remains uncertain, evidence suggests atypical femoral fractures are likely to be stress fractures, related to impaired repairing of microcracks in the bone. It is postulated that bisphosphonates may localize in areas where the stress fractures develop, preventing intracortical healing at the site of the atypical femoral fracture [35]. Atypical femoral fractures are uncommon, ranging from 3.0 to 9.8 cases per 100,000 patient years. However, the risk of atypical femoral fractures increases in patients with prolonged bisphosphonate use (more than 3–5 years). Contralateral involvement has been noted in approximately one-fourth of patients [37]. Clinicians should be aware of prodromal hip pain reported in 32–76% of patients, occurring two weeks to several years prefracture, and consider bilateral hip radiographs while evaluating the contralateral hip for symptoms and possible signs of a stress fracture. Bisphosphonates should be stopped if radiographic signs of atypical femoral fracture are noted. Consultation with an orthopedic surgeon should be obtained.

Patients without symptoms who are found to have an incomplete fracture can potentially be monitored off bisphosphonate therapy while avoiding high and repetitive impact activities. The first line of treatment for patients with a complete fracture is intramedullary nailing. Treatment for patients with hip pain and incomplete fractures is less certain, where conservative treatment or prophylactic intramedullary nailing can be considered based on consultation with orthopedic surgery [37].

7 Special Considerations

7.1 Chronic Kidney Disease

It is recommended to avoid bisphosphonate treatment in patients with reduced glomerular filtration rate (GFR) (<30 mL/min risedronate, ibandronate, and <35 mL/min alendronate and zoledronic acid) [4]. Oral bisphosphonates are not nephrotoxic, but this FDA recommendation reflects the lack of clinical trial data for patients with CKD, as studies excluded patients with creatinine levels of more than 2 mg/dL. Bisphosphonates have been shown to provide fracture benefits in patients with GFR as low as 15 mL/min in post-hoc analyses [38]. In patients with a low GFR, based on renal excretion of bisphosphonates, consideration could be given to reducing the dose of oral bisphosphonates if they are used. Early studies involving intravenous bisphosphonates (etidronate, clodronate) have shown that rapid infusions leading to higher levels of peak serum concentration lead to nephrotoxic effects on tubular cells [38]. Therefore, it is recommended that intravenous zoledronic acid be administered in no less than 15 min in patients with normal kidney function. In patients with a GFR of 35–50, consideration should be given to infusing intravenous zoledronic acid more slowly (30–60 min) [38].

8 Long-Term Use of Bisphosphonates

Due to the increased risk of atypical femoral fractures and possibly increased risk of osteonecrosis of the jaw with prolonged exposure to bisphosphonates, the duration of bisphosphonate treatment has become an area of focus. The Fracture Intervention Trial Long-term Extension (FLEX) trial [39] and the HORIZON-PFT trial [40] provide critical information about the benefits and risks of prolonged bisphosphonate treatment.

8.1 Alendronate

The FLEX trial [39] was a continuation of the FIT trial [12, 13]. The FLEX trial included all patients from the FIT trial [12, 13] who had received five years of alendronate. One thousand ninety-nine patients were then randomized to receive alendronate for ten years (5 mg daily or 10 mg daily) or placebo for an additional five years [39]. The effects of alendronate 5 and 10 mg daily were indistinguishable, and in a pooled analysis of the alendronate versus placebo group, lumbar spine BMD was maintained in both groups. Hip BMD decreased significantly in the placebo group compared to the alendronate extension group. In the alendronate extension group, clinical vertebral fractures were reduced by 55% ($p = 0.013$), while there were no differences in vertebral morphometric fractures and non-vertebral fractures. Post-hoc analyses showed a 50% relative risk reduction (95% Confidence Interval 0.26–0.96) of non-vertebral fracture risk in patients with a T-score ≤ -2.5 at FLEX baseline [41].

8.2 Zoledronic Acid

In the Extension to HORIZON-PFT trial, 1233 subjects who received zoledronic acid in the first three years were randomized to receive either placebo or three more years of intravenous zoledronic acid [40]. While vertebral BMD was the same, those receiving additional doses of intravenous zoledronic acid demonstrated increased BMD by 1.4% BMD compared to the placebo group. Morphometric vertebral fractures were reduced by 52% in those receiving the three additional years of zoledronic acid ($p < 0.0001$), while there were no differences between clinical and nonvertebral fractures.

8.3 Practical Considerations for Long-Term Bisphosphonate Use

For some patients, there is a certain degree of benefit with more prolonged use of bisphosphonates, especially in those specifically at higher risk of fracture, and particularly those at risk for vertebral fracture. After five years of alendronate or three years of zoledronic acid treatment, fracture risk should be reassessed. The treatment decision for ongoing bone health management should be individualized based on the patient's fracture risk, other treatment options, and preferences. For patients at lower risk for fracture, a bisphosphonate drug holiday should be considered. Patients with a higher risk of fracture: age 70–75 years, T-score ≤-2.5, or FRAX (Fracture Risk Assessment Tool) risk score that is above country-specific thresholds, prior osteoporotic fracture, or patients sustain a fracture while taking osteoporosis treatment may benefit from a different approach.

One could consider continuation of treatment for up to 10 years for oral bisphosphonates or up to 6 years for intravenous zoledronic acid for those patients at particularly high risk for fracture. Another option would be to consider switching to an alternative osteoporosis treatment such as denosumab, an osteoanabolic agent-teriparatide, abaloparatide, or romosozumab [42]. During a bisphosphonate holiday, continued careful and close monitoring with DXA scan and regular reassessment of the patient's fall risk and general medical health (i.e. initiation of medications such as glucocorticoids or aromatase inhibitors) is recommended.

Some clinicians monitor bone turnover markers during the drug holiday. Of note, bone turnover markers are biologically variable, and should be performed as a fasting specimen, first thing in the morning. One would consider resuming treatment for osteoporosis when significant bone loss is noted on DXA scan or when there is a substantial increase in bone turnover markers. With conclusion of the drug holiday, it is reasonable to resume any of the guideline-recommended treatment regimens, and this could include a bisphosphonate.

9 Summary

With the burden of osteoporosis and osteoporotic fractures expected to increase in the coming decades, bisphosphonates are likely to continue to be cornerstones in bone health management due to their efficacy and cost-effectiveness. While there are potential adverse events with bisphosphonate treatment, such as the rare occurrences of atypical femoral fractures or osteonecrosis of the jaw that have raised concern about their use, it is essential to maintain perspective that these associations are very rare. The benefit of these medications in reducing the risk of fragility fractures still significantly outweighs the small potential risks associated with treatment. Future research is needed to further refine the use of these highly efficacious medications, especially in terms of optimal duration of treatment and duration of drug holiday, to maintain maximum benefit while minimizing any potential, albeit small, risk to patients with osteoporosis.

References

1. Lim SY, Bolster MB (2015) Current approaches to osteoporosis treatment. Curr Opin Rheumatol 27:216–224
2. Fleisch H, Russell RG, Francis MD (1969) Diphosphonates inhibit hydroxyapatite dissolution in vitro and bone resorption in tissue culture and in vivo. Science 165:1262–1264
3. Drake MT, Clarke BL, Khosla S (2008) Bisphosphonates: mechanism of action and role in clinical practice. Mayo Clin Proc 83:1032–1045
4. Diab DL, Watts NB, Miller PD (2021) Bisphosphonates pharmacology and use in the treatment of osteoporosis. In: Dempster DW, Cauley JA, Bouxsein ML et al (eds) Marcus and Feldman's osteoporosis, 5th edn. Academic Press, Massachusetts, pp 1721–1736
5. Lin JH (1996) Bisphosphonates: a review of their pharmacokinetic properties. Bone 18:75–85
6. Russell RG (2006) Bisphosphonates: from bench to bedside. Ann N Y Acad Sci 1068:367–401
7. Roelofs AJ, Ebetino FH, Reszka AA et al (2008) Bisphosphonates: mechanisms of action. In: Bilezikian JP, Raisz LG, John Martin T (eds) Principles of bone biology, 3rd edn. Academic Press, Massachusetts, pp 1737–1767
8. Wiesner A, Szuta M, Galanty A et al (2021) Optimal dosing regimen of osteoporosis drugs in relation to food intake as the key for the enhancement of the treatment effectiveness-a concise literature review. Foods 10:720. https://doi.org/10.3390/foods10040720
9. Cremers S, Drake MT, Ebetino FH et al (2020) Clinical and translational pharmacology of bisphosphonates. In: Bilezikian JP, John Martin T, Clemens TL et al (eds) Principles of bone biology, 4th edn. Academic Press, Massachusetts, pp 1671–1687
10. Black DM, Delmas PD, Eastell R et al (2007) Once-yearly zoledronic acid for treatment of postmenopausal osteoporosis. N Engl J Med 356:1809–1822
11. McClung MR (2021) Role of bone-forming agents in the management of osteoporosis. Aging Clin Exp Res 33:775–791
12. Black DM, Thompson DE, Bauer DC et al (2000) Fracture risk reduction with alendronate in women with osteoporosis: the Fracture Intervention Trial. FIT Research Group. J Clin Endocrinol Metab 85:4118–4124
13. Cummings SR, Black DM, Thompson DE et al (1998) Effect of alendronate on risk of fracture in women with low bone density but without vertebral fractures: results from the Fracture Intervention Trial. JAMA 280:2077–2082.
14. Harris ST, Watts NB, Genant HK et al (1999) Effects of risedronate treatment on vertebral and nonvertebral fractures in women with postmenopausal osteoporosis: a randomized controlled trial. Vertebral efficacy with risedronate therapy (VERT) study group. JAMA 282:1344–1352
15. Reginster J, Minne HW, Sorensen OH et al (2000) Randomized trial of the effects of risedronate on vertebral fractures in women with established postmenopausal osteoporosis. Vertebral efficacy with risedronate therapy (VERT) study group. Osteoporos Int 11:83–91
16. McClung MR, Geusens P, Miller PD et al (2001) Effect of risedronate on the risk of hip fracture in elderly women. Hip intervention program study group. N Engl J Med 344:333–340
17. Chesnut CH 3rd, Skag A, Christiansen C et al (2004) Effects of oral ibandronate administered daily or intermittently on fracture risk in postmenopausal osteoporosis. J Bone Miner Res 19:1241–1249
18. Lyles KW, Colón-Emeric CS, Magaziner JS et al (2007) Zoledronic acid and clinical fractures and mortality after hip fracture. N Engl J Med 357:1799–1809
19. Harold R (2022) Bisphosphonate therapy for the treatment of osteoporosis. https://www.uptodate.com/contents/bisphosphonate-therapy-for-the-treatment-of-osteoporosis. Accessed 31 March 2023

20. Colón-Emeric CS, Mesenbrink P, Lyles KW et al (2010) Potential mediators of the mortality reduction with zoledronic acid after hip fracture. J Bone Miner Res 25:91–97
21. Schnitzer T, Bone HG, Crepaldi G et al (2000) Therapeutic equivalence of alendronate 70 mg once-weekly and alendronate 10 mg daily in the treatment of osteoporosis. Alendronate once-weekly study group. Aging 12:1–12
22. Brown JP, Kendler DL, McClung MR et al (2002) The efficacy and tolerability of risedronate once a week for the treatment of postmenopausal osteoporosis. Calcif Tissue Int 71:103–111
23. Delmas PD, McClung MR, Zanchetta JR et al (2008) Efficacy and safety of risedronate 150 mg once a month in the treatment of postmenopausal osteoporosis. Bone 42:36–42
24. Reginster JY, Adami S, Lakatos P et al (2006) Efficacy and tolerability of once-monthly oral ibandronate in postmenopausal osteoporosis: 2 year results from the MOBILE study. Ann Rheum Dis 65:654–661
25. Delmas PD, Adami S, Strugala C et al (2006) Intravenous ibandronate injections in postmenopausal women with osteoporosis: one-year results from the dosing intravenous administration study. Arthritis Rheum 54:1838–1846
26. Woo C, Gao G, Wade S et al (2010) Gastrointestinal side effects in postmenopausal women using osteoporosis therapy: 1-year findings in the POSSIBLE US study. Curr Med Res Opin 26:1003–1009
27. Cryer B, Bauer DC (2002) Oral bisphosphonates and upper gastrointestinal tract problems: what is the evidence? Mayo Clin Proc 77:1031–1043
28. Federal Drug Administration (2012) Fosamax (alendronate sodium) tablets label. https://www.accessdata.fda.gov/drugsatfda_docs/label/2012/021575s017lbl.pdf. Accessed 31 March 2023
29. de Groen PC, Lubbe DF, Hirsch LJ et al (1996) Esophagitis associated with the use of alendronate. N Engl J Med 335:1016–1021
30. Baker DE (2002) Alendronate and risedronate: what you need to know about their upper gastrointestinal tract toxicity. Rev Gastroenterol Disord 2:20–33
31. Lim SY, Bolster MB (2018) What can we do about musculoskeletal pain from bisphosphonates? Cleve Clin J Med 85:675–678
32. Khan AA, Morrison A, Hanley DA et al (2015) Diagnosis and management of osteonecrosis of the jaw: a systematic review and international consensus. J Bone Miner Res 30:3–23
33. Wan JT, Sheeley DM, Somerman MJ et al (2020) Mitigating osteonecrosis of the jaw (ONJ) through preventive dental care and understanding of risk factors. Bone Res 8:14. https://doi.org/10.1038/s41413-020-0088-1
34. Ruggiero SL, Dodson TB, Fantasia J et al (2014) American association of oral and maxillofacial surgeons position paper on medication-related osteonecrosis of the jaw–2014 update. J Oral Maxillofac Surg 72:1938–1956
35. Shane E, Burr D, Abrahamsen B et al (2014) Atypical subtrochanteric and diaphyseal femoral fractures: second report of a task force of the American society for bone and mineral research. J Bone Miner Res 29:1–23
36. Black DM, Geiger EJ, Eastell R et al (2020) Atypical femur fracture risk versus fragility fracture prevention with bisphosphonates. N Engl J Med 383:743–753
37. Silverman S, Kupperman E, Bukata S (2018) Bisphosphonate-related atypical femoral fracture: managing a rare but serious complication. Cleve Clin J Med 85:885–893
38. Miller PD (2011) The kidney and bisphosphonates. Bone 49:77–81
39. Black DM, Schwartz AV, Ensrud KE et al (2006) Effects of continuing or stopping alendronate after 5 years of treatment: the fracture intervention trial long-term extension (FLEX): a randomized trial. JAMA 296:2927–2938

40. Black DM, Reid IR, Boonen S et al (2012) The effect of 3 versus 6 years of zoledronic acid treatment of osteoporosis: a randomized extension to the HORIZON-pivotal fracture trial (PFT). J Bone Miner Res 27:243–254

41. Schwartz AV, Bauer DC, Cummings SR et al (2010) Efficacy of continued alendronate for fractures in women with and without prevalent vertebral fracture: the FLEX trial. J Bone Miner Res 25:976–982

42. Adler RA, El-Hajj Fuleihan G, Bauer DC et al (2016) Managing osteoporosis in patients on long-term bisphosphonate treatment: report of a task force of the American society for bone and mineral research. J Bone Miner Res 31:16–35

Denosumab: Clinical Applications, Outcomes, and Perspectives in Osteoporosis

Nouran Eshak, Afrina Rimu, and Alexandra Hoffman

Abstract Increased understanding of the pathways involved in bone metabolism has led to the development of highly specific biologic medications that can be used in the treatment of osteoporosis. In this chapter, we will review the clinical uses of denosumab, a fully human IgG2 monoclonal antibody which regulates the bone remodeling pathway. We will provide the reader a basic understanding of the properties, safety, and efficacy of this monoclonal antibody and its clinical applications in the prevention and treatment of osteoporosis and other skeletal related events.

Keywords Denosumab · Postmenopausal osteoporosis · Skeletal related events · Glucocorticoid induced osteoporosis

1 Introduction

The RANKL/RANK/OPG system is an important pathway in the regulation of bone remodeling and one which is targeted by the osteoporosis drug, denosumab. NF-κ-β ligand (RANKL) is a glycoprotein expressed by osteoblasts and stromal cells which regulates the activity of osteoclasts including their differentiation, activation, and survival. RANKL binds the RANK receptors on osteoclasts leading to their activation and subsequent bone resorption [1].

In the 1990s, with the discovery of this pathway, it was noted that osteoprotegerin (OPG), or osteoclastogenesis inhibitory factor, prevents the binding of RANKL to RANK thereby reducing osteoclast formation and bone turnover [2]. This led to a desire for the development of an anti-RANKL antibody which was accomplished when denosumab was approved for the treatment of osteoporosis in postmenopausal women in 2010.

N. Eshak (✉)
Rheumatology Department, Mayo Clinic Arizona, Scottsdale, Az, USA
e-mail: nouran.eshak@ttuhsc.edu

A. Rimu · A. Hoffman
Internal Medicine, Texas Tech University Health Sciences Center, Lubbock, TX, USA

Fig. 1 Summary of the action of denosumab. Denosumab acts similarly to osteopetrogen a naturally occurring decoy receptor of RANKL preventing the binding of RANKL to RANK receptors on osteoclasts and their precursors, leading to impaired osteoclast differentiation and function

Denosumab is a fully human IgG2 monoclonal antibody that binds RANKL with high affinity and specificity in both its soluble and membrane-bound forms preventing activation of the RANKL/RANK pathway and subsequent bone resorption leading to increased bone density (Fig. 1). Denosumab has a rapid onset and offset of action with bone turnover markers returning to pretreatment levels within 9 months of treatment cessation and is fully reversible [3], unlike bisphosphonates which become embedded in bone tissue, denosumab is thought to be cleared by the reticuloendothelial system with a half-life of approximately 26 days. With convenient twice-yearly dosing and a novel mechanism of action, denosumab has quickly become a common and well-tolerated drug for the treatment of osteoporosis [1]. This chapter aims to review denosumab's efficacy in the treatment of osteoporosis in different patient populations.

2 Clinical Applications of Denosumab

2.1 Denosumab in Treatment of Osteoporosis in Postmenopausal Women

Approximately 16.5% of American women above the age of 65 have osteoporosis, defined by a T-score of <-2.5 at the femoral neck or lumbar spine.

Osteoporosis is a disease characterized by low bone mass and microarchitectural disturbances leading to bone fragility and increase risk of fractures. In postmenopausal women the lack of estrogen leads to an increase RANKL effect, leading to higher bone resorption and loss of bone mass with eventual osteoporosis.

Denosumab a RANKL antagonist inhibits osteoclast function leading to higher bone mass. The FREEDOM Trial was a pivotal phase III clinical trial on denosumab for the treatment of osteoporosis in postmenopausal women. It was a placebo-controlled trial extending for 36 months with denosumab given at 60 mg q 6 months. The primary endpoint was the prevention of vertebral fractures. The trial showed a reduction of vertebral fractures by 68%, hip by 40% and non-vertebral by 20%, without increase in the incidence of adverse events in the first 36 months. In addition, it showed an increase in BMD at the lumbar spine (LS) and hip by 9 and 6%, respectively. There was no noted increase in incidence of adverse events (AE), including jaw osteonecrosis or hypocalcemia [4].

The FREEDOM Trial was extended for another 7 years with the main aim on collecting safety and efficacy data with long-term denosumab use. Those in the denosumab group continued it (long-term group), and the ones that were on placebo were switched to denosumab (cross over group). Denosumab showed continued efficacy with continued low fracture incidence and continuous increase in BMD with hip BMD increasing by 9–10% and LS by about 20%. The annual rates of vertebral and non-vertebral fractures remained low [5]. The BMD gain does not plateau with denosumab use as opposed to other osteoanabolic and anti-resorptive medications, because although denosumab suppresses bone resorption it may continue to allow a modeling effect on the bone [6].

Two other phase III clinical trials were head-to-head trials on denosumab versus alendronate in post-menopausal women with the end point being increase in BMD.

The DECIDE trial was conducted on women with low BMD who have not been on bisphosphonates for the past year, while the STAND trial included women previously on alendronate to assess the safety of transitioning from alendronate to denosumab. Both trials showed similar results with higher gains on BMD at all sites in the denosumab group versus the alendronate group, with similar rates of adverse events. More fractures occurred in the denosumab group; however, these results did not reach statistical significance, as the trials were not powered to assess fracture rates [7, 8].

2.2 Denosumab Safety

When given at the dose of 60 mg subcutaneous every 6 months, denosumab has a good safety profile.

In the FREEDOM trial, there was a higher rate of cellulitis among patients in the denosumab group than the placebo. And though there was a higher rate on infection related serious adverse events, it did not reach statistical significance. Eczema also occurred more frequently among the denosumab group. During the 7-year extension phase of the trial, incidence adverse events continued to be stable or declining, overall decreasing from 165 to 95.9 events per 100 participant years. Rates of hypocalcemia continued to be very low <0.1 per 100 participant years; similarly, in the DECIDE trial there were no reports of symptomatic hypocalcemia. Two atypical femoral fractures occurred, one in the long term and one in the cross over group with a rate 0.8/10000 participant years. Thirteen cases of jaw osteonecrosis occurred, frequency of 5.2/10 000 participant years, 7 in the long term and 6 in the cross over group, 2 patients withdrew from the trial, the remaining 11 patients all had complete resolution of jaw osteonecrosis, 4 while still on denosumab therapy.

2.3   Denosumab in Clinical Practice

The North American Menopause Society statement stratifies osteoporosis treatment according to a patient's risk of fracture. In post-menopausal women with low risk of fractures it would be reasonable to start with raloxifene or bisphosphonates, while in those with a high risk, bisphosphonates or denosumab should be considered first choices of therapy. For patients with a very high risk of fractures e.g. very low T-scores, or previous fragility fractures, an osteo-anabolic agent may be considered as these produce faster and higher bone gain, and this should then be followed by an anti-remodeling drug as bisphosphonate or denosumab, to retain the bone [9].

The most up to date Endocrine Society treatment guidelines recommend bisphosphonate or denosumab as the first line pharmacological treatment for post-menopausal osteoporosis. Denosumab should be administered at the dose of 60 mg q 6 months, and fracture risk should be assessed after 5–10 years [10].

Patients on denosumab must either continue the drug indefinitely or be transitioned to another medication. A drug holiday should be avoided. A post-hoc analysis on the FREEDOM trial and extension showed that patients who discontinue denosumab have an increase in bone turnover markers 3 months after the omitted dose, and a regression of BMD to baseline 6 months after omitted dose. In addition, there was an increase in vertebral fracture rate from 1.2 to 7.1% similar to those who were on placebo, with an increased risk of multiple vertebral fractures 3.4% in those who discontinued denosumab versus 2.2% in those on placebo, especially in patients who had a previous vertebral fracture OR (3.9) [11].

2.4   Denosumab in Chronic Kidney Disease

Denosumab is mainly metabolized through the reticuloendothelial system with almost no renal clearance. Consequently, unlike bisphosphonates which are cleared by the kidneys, renal impairment does not affect the pharmacokinetics or pharmacodynamics of denosumab, and this drug can be given in patients with CKD with no dose adjustments [12].

Patients with mild to moderate CKD (i.e., eGFR >30 ml) with no evidence of mineral bone disease do not usually develop adverse effects from denosumab.

Patients with advanced CKD and ESRD may have co-existing osteoporosis and mineral bone disease or renal osteodystrophy. Although DEXA scan is not the ideal investigation to differentiate between these diseases, a low bone mineral density is still a strong predictor of fragility fractures in patients with CKD. Evidence of mineral bone disease includes hypocalcemia, hyperphosphatemia, secondary hyper-parathyroidism, and vitamin D deficiency. Care must be taken to identify other forms of renal bone disease prior to initiating treatment with denosumab, as it may cause severe and prolonged hypocalcaemia [13].

In a meta-analysis of observational studies on denosumab use in patients with ESRD, hypocalcaemia rate was 42%, occurring 7–20 days after administration, but was associated with decrease in parathyroid hormone levels and increase in bone mineral density [14].

Another study reported that the mean time to nadir of calcium was 21 days, and the mean time to correction was 71 days, correction usually required high doses of calcium, calcitriol and increased calcium concentration in the dialysate [15].

Prior to initiating treatment, it is important to discuss risk versus benefit with the patient, diagnose and treat mineral bone disease by optimizing calcium, phos-phorus and vitamin D levels, closely monitor calcium levels after administration, and continue calcium and vitamin D supplementation.

2.5 Denosumab in Prevention and Treatment of Osteoporosis in Women Receiving Aromatase Inhibitors

Patients with breast cancer being treated with aromatase inhibitors are at increased risk to develop osteoporosis. In postmenopausal women, the residual production of estrogen is maintained through the conversion of adrenal androgen precursors into estrogen in peripheral tissues by the enzyme CYP19A1, or aromatase. This enzyme is highly expressed in adipose and breast tissue, and its elevated activity in breast cancer stimulates the growth of estrogen receptor-positive malignant cells making it an excellent target for endocrine adjuvant therapy. Due to the development of aromatase inhibitors, such as anastrozole, exemestane, and letrozole, estrogen receptor-positive breast cancer is associated with a favorable prognosis with increased survival and subsequently longer lifespan. For this reason, adjuvant endocrine therapy is the treatment of choice for hormone receptor-positive early-stage breast cancer [16]. However, as these treatments affect systemic estrogen levels, they potentially decrease bone density.

Estrogens decrease the activity of the RANKL/RANK/OPG pathway leading to suppression of bone resorption by decreasing RANKL, increasing OPG, and desen-sitizing RANK, thereby inducing osteoclast apoptosis and inhibiting osteoclasto-genesis. For this reason, the relative estrogen deficiency produced via aromatase blockade leads to increased bone resorption due to increased activity of the RANK pathway, and aromatase inhibitor therapy has been shown to lead to bone loss in several studies [17]. Therefore, the benefit these medications confer in breast cancer comes at the cost of bone density. Maintenance of bone health in long-term breast cancer survivors then becomes clinically challenging and with breast cancer being the most prevalent malignancy in females with 1 in 8 developing breast cancer at some point in their lifetime, it is a very clinically relevant challenge [17]. However, one solution to the problem of bone loss in aromatase inhibitor therapy is denosumab.

In 2009, one clinical trial randomized hormone receptor-positive breast cancer patients with a decreased bone mass on aromatase inhibitors to denosumab or placebo. After 12 and 24 months the denosumab group had increased lumbar spine bone density (5.5% and 7.6%, respectively, $p < 0.0001$ at both time points) compared to a placebo. Increases in bone density were observed as early as one month and were not affected by the duration of aromatase inhibitor therapy [18].

A subsequent multicenter, randomized, double-blind placebo-controlled trial, ABCSG-18, investigated the effects of denosumab in postmenopausal women with early-stage hormone receptor-positive breast cancer being treated with aromatase inhibitors. 3420 patients were randomized to denosumab or placebo and treated until the prespecified number of 247 fractures was achieved. The addition of denosumab to aromatase inhibitor therapy reduced the risk of clinical fractures, and the first fracture for the denosumab group was significantly delayed (HR 0.50, 95% CI 0.39–0.65, $P < 0.0001$). Additionally, the denosumab group had fewer fractures overall (92 compared to 176); these results were independent of baseline T score or age. The overall incidence of treatment-related adverse events was similar between treatment groups [16].

2.6   Denosumab in Treatment of Men with Osteoporosis

Osteoporosis in men may be either primary i.e., idiopathic, or secondary, which is more common in men than women, with the most common causes being alcohol use, glucocorticoid excess which is usually exogenous, and hypogonadism. Hypogonadism can either be primary or induced, e.g., in patients with prostate cancer taking androgen deprivation therapy. Although men are less commonly affected by osteoporosis than women, they have a higher mortality rate following a hip fracture than women, with 38% versus 28% 1-year mortality and 10% in-hospital mortality. Other osteoporotic fractures such as vertebral fractures also have increased mortality [19]. Denosumab safety and efficacy in men with osteoporosis were studied in the ADAMO trial which included 242 men randomized to receive either denosumab for 1 year or placebo followed by 1 year where all patients received denosumab. Long-term denosumab treatment was associated with cumulative gain in 24-month bone mineral density (BMD) from baseline (8.0% at the lumbar spine, 3.4% total hip, 3.4% femoral neck, 4.6% trochanter, and 0.7% at the radius). Similar BMD gains after 12 months of denosumab treatment were observed in the crossover group. However, the trial did not show any benefits in terms of fracture risks as fracture rates were equivalent in both groups [20].

The DIRECT trial studied fracture risk reduction with denosumab in Japanese postmenopausal women and men with osteoporosis in a double-blind placebo-controlled design. The primary endpoint was the 24-month incidence of new or worsening vertebral fracture for denosumab versus placebo. A subgroup analysis of men included 23 and 24 men in denosumab and placebo groups, respectively, and demonstrated an incidence of new or worsening vertebral fractures of 0 and 12.5% in the denosumab and placebo groups respectively while the incidence of new fractures were 0% and 8.3% in the denosumab and placebo groups, respectively. However, these results did not reach statistical significance [21].

3 Denosumab Use in Men with Prostatic Cancer

3.1 Androgen Deprivation Therapy Related-Osteopenia and Osteoporosis

Prostate cancer accounts for 25% of annual cancer diagnoses; treatment for metastatic prostate cancer and some cases of non-metastatic cancer include androgen deprivation therapy (ADT) using gonadotropin hormone agonists. This results in a rapid decline in the circulating sex hormones leading to osteoclast activation and osteoblast apoptosis with consequent bone resorption resulting in a 5–10% annual loss in bone mass as opposed to the normal annual loss of 0.5–1%. Loss of bone mass along with ADT-induced sarcopenia, increases the risk of falls, and fractures, with consequent loss of function and increased morbidity. With prolonged survival rates for prostate cancer, bone health is becoming an emerging health concern [22]. Lifestyle modifications, exercise, and calcium/vitamin D intake should be reinforced in all patients on ADT. FRAX score including a bone mineral density if available, should be calculated. Treatment with anti-resorptive therapy should be initiated if the treatment threshold is met, if not, comprehensive reevaluation should be repeated in 12–18 months [23]. Denosumab and bisphosphonates are currently the only approved treatments in patients with prostate cancer. Teriparatide is contraindicated in patients with bone metastasis or previous radiotherapy, and selective estrogen receptor modulators are currently not recommended to prevent bone loss in patients on ADT. Trials conducted on bisphosphonates did not show a statistically significant reduction in the incidence of fractures, however, a meta-analysis that included 2634 patients demonstrated that bisphosphonates had a substantial effect in preventing fractures and bone loss prevention was reached without major side effects (cardiovascular or gastrointestinal events) [24].

In a large, randomized placebo-controlled trial, 60 mg of denosumab for 6 months in patients with non-metastatic prostate cancer was associated with an increase in BMD at the lumbar spine, femoral neck, hip, and radius at 2 and 36 months, with a lower incidence of vertebral fractures at 1.5% versus 3.9% compared to placebo. Only 14.7% had osteoporosis with a T score of less than −2.5 [25]. Another trial studied 60 mg of denosumab for 6 months versus 70 mg of alendronate weekly in osteoporosis patients with prostate cancer on ADT. Although vertebral fractures' incidence was lower with denosumab, it did not reach a statistical significance; however BMD at the lumbar spine was significantly improved in patients on denosumab compared to alendronate, 5.5% versus −1.1% [26].

3.2 Prevention of Skeletal-Related Events in Prostate Cancer

For prevention of skeletal-related events in patients with metastatic prostate cancer, a monthly dose of 120 mg of denosumab was found to be superior to a monthly dose of 4 mg of Zoledronic acid, prolonging time to the first event; however, there was no benefit in overall survival or disease progression with higher rates of hypocalcemia [27].

3.3 Prevention of Bone Metastasis in Prostate Cancer

Preclinical models of prostate cancer showed that osteoclast inhibition may prevent bone metastasis; one theory is that RANKL may act as a hormonal signal to tumor cells and that changing the bone microenvironment may prevent bone metastasis. Denosumab was compared to a placebo for bone metastases free survival in castration-resistant prostate cancer confirming that there was an increase in time to first bone metastasis by 4.2 months; however, adverse effects, such as hypocalcemia and jaw osteonecrosis, were expectedly more frequent in the denosumab group, though similar to rates in other studies using same denosumab dosage [28].

3.4 Denosumab Use in Glucocorticoid Induced Osteoporosis

About 1% of U.S. population is treated with some forms of glucocorticoid (GC) for various inflammatory conditions [29]. Even at low-doses, GC can cause osteoporosis, bone loss and fractures in the long run. Calcium, vitamin D and bisphosphonates (alendronate, risedronate, zoledronic acid), teriparatide all are recommended as preventive measures of glucocorticoid induced osteoporosis (GIO) [30–34]. Side effects caused by bisphosphonates and limited duration of treatment recommendations by teriparatide act as potential barriers for their use [35, 36]. The monoclonal antibody, denosumab, has the potential to overcome these problems for GIO [37]. However, recommendations by American college of Rheumatology [38] for GIO has given less emphasis on the use of denosumab, because of lack of safety data and only recommended for patients ≥40 years old with moderate or high risk of fractures after other therapies (oral and intravenous bisphosphonates, Teriparatide) have been tried [38]. The updated guidelines from 2022 suggested denosumab as conditional recommendations in moderate and high-risk fractures in GIO [39].

Several clinical trials have been conducted to see the effect of denosumab in patients with osteoporosis who have histories of long-time use of GC, most of them showed promising results. A randomized trial by Mok et al. in 2015 compared the continued use of oral bisphosphonates with switching of oral bisphosphonates to denosumab in adult patients with long-term GC therapy (2 years) [40]. Denosumab group had higher increase in BMD in hip and spine compared to continuing bisphosphonates group ($+3.4 \pm 0.9\%$ ($p = 0.002$) vs. $+1.5 \pm 0.4\%$ ($p = 0.001$) in hip and $+1.4 \pm 0.6\%$ ($p = 0.03$) vs. $+0.80 \pm 0.5\%$ ($p = 0.12$) in spine, respectively) and stronger suppression of bone turnover markers at 12-month. Only minor infections were more prevalent in denosumab group.

A single center randomized controlled trial by Iseri et al. on 28 glomerular disease patients receiving GC (prednisolone) who were also diagnosed with GIO showed similar positive increase in BMD at lumbar spine both at 6 months and 12 months (6 months: $+2.9\% \pm 0.7\%$, $p < 0.05$; 12 months: $+5.3\% \pm 1.0\%$, $p < 0.01$), whereas alendronate treatment failed to show any significant increase in BMD at lumbar spine, femoral neck etc. both at 6 months and 12 months, although the latter was relatively safer [41].

A large-scale, randomized, double blind, multicenter (79 centers in Europe, Latin America, Asia, North America) trial with 795 patients by Saag et al. compared Denosumab to Risedronate in GC continuing (≥ 3 months) and initiating (<3 months) patients who had osteoporosis related fractures [37, 42]. Denosumab was found out to be non-inferior and superior compared to Risedronate in both GC continuing (4.4% [95% CI 3.8–5.0] vs. 2.3% [1.7–2.9]; $p < 0.0001$) and initiating group (3.8% [3.1–4.5] vs. 0.8% [0.2–1.5]; $p < 0.0001$) in regards to mean percent increase in BMD, increase in lumbar spine, total hip, femoral neck, 1/3 radius in GC initiating group. For continuing group, a similar positive result was found. Safety profile was similar between the two groups. A sub-study of the above-mentioned trial conducted to assess the bone strength and trabecular microarchitecture by standard analyses and failure load (FL) with micro-finite element analysis showed that FL at radius remained unchanged in Denosumab, even decreased in Risedronate group for the initiating group at 24-month [43]. For the continuing group, FL increased in Denosumab but not in Risedronate group.

A case–control study conducted by Matsuno et al. on postmenopausal female patients with rheumatoid arthritis who had long-term GC therapy and osteoporosis reported similar findings of increase in BMD after denosumab use [44]. Compared to bisphosphonates, patients receiving denosumab showed statistically significant improvement at BMD over 2-year period. No notable adverse effects were observed in this GC taking group after taking denosumab.

The above-mentioned study results conducted in the last few years corroborate the finding that denosumab can be deemed as a potential therapeutic option for treatment and prevention of GIO due to its relative safety profile, and its convenience of dosing every 6-month subcutaneously.

4 Other Indications for Denosumab Use

4.1 Denosumab in Hypercalcemia of Malignancy

Tumor-induced, osteoclast-mediated bone resorption and resulting hypercalcemia, commonly known as hypercalcemia of malignancy (HCM) is very common in advanced cancers, such as prostate cancer, cervical cancer, multiple myeloma, squamous malignancy of head and neck, carcinoma of breast, etc. [45, 46]. Parathyroid hormone-related protein (PTHrP) secreted by the malignant cells is responsible for the very high levels of calcium in blood and causes symptoms of hypercalcemia such as nausea, vomiting, abdominal pain, bone pain, fatigue, pathological fractures. Grave complications of HCM include renal failure, coma and ultimately death [45].

Bisphosphonate therapy is the treatment of choice for HCM [47]. In multiple studies, use of bisphosphonates, like zoledronic acid (ZA) and pamidronate, caused normalization of serum calcium (≤ 10.8 mg/dl) in about 88% and 70% of patients, respectively [48, 49]. Very recently, denosumab use has also shown promising results in the treatment of HCM. Multiple studies have already corroborated the findings of denosumab to delay the initiation of skeletal-related events (SRE) and eventual hypercalcemia in patients with advanced cancers. One of the important adverse events in these studies were hypocalcemia due to denosumab use, which can potentially be seen as a treatment strategy for HCM [50, 51].

A single arm study conducted by Hu et al. investigated the role of denosumab use in advanced cancer patients who failed bisphosphonate therapy initially. Subcutaneous denosumab (120 mg) use on days 1, 8, 15 and 29 and then q 4 weeks had a favorable response, with the serum calcium level reaching to ≤ 11.5 mg/dl in majority of the patients at day 10 and 21. Median duration of 104 days was required to achieve the reduction in serum calcium in this study cohort [45]. Another phase III, randomized trial in patients with breast, other solid cancers and multiple myeloma by Diel et al. to compare denosumab to bisphosphonate reported a statistically significant delay in the onset of HCM with denosumab use compared to ZA use (hazard ratio 0.63) and reduction in recurrent HCM [52]. Further comparative studies are warranted to determine the efficacy of denosumab in prevention and treatment of HCM.

4.2 Denosumab Use in Multiple Myeloma

Multiple myeloma (MM) is an aggressive cancer originating from plasma cells in the bone marrow and is the second most common hematological malignancy overall [53]. Skeletal-related events (SRE), such as osteoclast mediated bone resorption, which can eventually lead to fractures, is one of the important manifestations of MM. As many as 80–90% of patients with MM can suffer from some form of SRE during their disease course. Bisphosphonates have been recommended as the overall preventive treatment for SRE in MM [50]. It has been shown that compared to placebo, zoledronic acid

(ZA) prolongs the time for the development of SRE in MM.54 However, various side effects of bisphosphonates (for example, renal complications) have limited the use of bisphosphonate in patients with MM.

Recent trials have shown promising results with using denosumab for SRE in MM patients. A study on Asian population conducted by Huang et al. supported the additional use of Denosumab in the treatment of lytic bone lesions in MM patients, with fewer patients taking denosumab in the study having SRE compared ZA group. Adverse effects, like jaw osteonecrosis and hypocalcemia, were comparable between the two groups [54]. Denosumab was also found to be non-inferior in delaying time for SRE development in two different studies (hazard ratio of 0.84 and 0.98, respectively) [55], although not found to be superior compared to ZA. Both groups exhibited similar progression of disease course [50, 51]. In a sub-group analysis, the latter group (Raje et al.) found that the net monetary benefit of US$10,259 with improvement of quality of life for the patients with denosumab group, making it more favorable compared to ZA [53].

In summary, although bisphosphonates are considered a first line options for prevention and treatment of SRE in MM, denosumab is gaining more recognition as an additional treatment due to its comparative safety profile, patient compliance and cost-effectiveness.

5 Conclusion

In this chapter, we described denosumab, a human monoclonal antibody used in the treatment of osteoporosis and its uses in various patient populations. The specificity of this drug allows for precise targeting of the pathways involved in bone remodeling, specifically the RANKL/RANK/OPG system. Because denosumab is generally well tolerated with a good safety profile and convenient twice-yearly dosing, it is has become instrumental in expanding the tools available for treatment of not only primary and secondary osteoporosis in both men and women, but also multiple myeloma and hypercalcemia of malignancy.

References

1. Hanley DA, Adachi JD, Bell A et al (2012) Denosumab: mechanism of action and clinical outcomes. Int J Clin Pract 66:1139–1146
2. Boyce BF, Xing L (2007) The RANKL/RANK/OPG pathway. Curr Osteoporos Rep 5:98–104
3. Lacey DL, Boyle WJ, Simonet WS et al (2012) Bench to bedside: elucidation of the OPG-RANK-RANKL pathway and the development of denosumab. Nat Rev Drug Discov 11:401–419
4. Cummings SR, San Martin J, McClung MR et al (2009) Denosumab for prevention of fractures in postmenopausal women with osteoporosis. N Engl J Med 361:756–765

5. Bone HG, Wagman RB, Brandi ML et al (2017) 10 years of denosumab treatment in post-menopausal women with osteoporosis: results from the phase 3 randomised FREEDOM trial and open-label extension. Lancet Diabetes Endocrinol 5:513–523
6. Ominsky MS, Libanati C, Niu QT et al (2015) Sustained modeling-based bone formation during adulthood in cynomolgus monkeys may contribute to continuous BMD gains with denosumab. J Bone Miner Res 30:1280–1289
7. Brown JP, Prince RL, Deal C al (2009) Comparison of the effect of denosumab and alendronate on BMD and biochemical markers of bone turnover in postmenopausal women with low bone mass: a randomized, blinded, phase 3 trial. J Bone Miner Res 24:153–161
8. Kendler DL, Roux C, Benhamou CL et al (2010) Effects of denosumab on bone mineral density and bone turnover in postmenopausal women transitioning from alendronate therapy. J Bone Miner Res 25:72–81
9. Management of Osteoporosis in Postmenopausal Women: The 2021 Position Statement of The North American Menopause Society" Editorial Panel (2021) Management of osteoporosis in postmenopausal women: the 2021 position statement of The North American Menopause Society. Menopause 28:973–997
10. Shoback D, Rosen CJ, Black DM et al (2020) Pharmacological management of osteoporosis in postmenopausal women: an endocrine society guideline update. J Clin Endocrinol Metab 105:dgaa048. https://doi.org/10.1210/clinem/dgaa048
11. Cummings SR, Ferrari S, Eastell R et al (2018) Vertebral fractures after discontinuation of denosumab: a post hoc analysis of the randomized placebo-controlled FREEDOM trial and its extension. J Bone Miner Res 33:190–198
12. Miyazaki T, Tokimura F, Tanaka S (2014) A review of denosumab for the treatment of osteoporosis. Patient Prefer Adherence 8:463–471
13. Nitta K, Yajima A, Tsuchiya K (2017) Management of osteoporosis in chronic kidney disease. Intern Med 56:3271–3276
14. Thongprayoon C, Acharya P, Acharya C et al (2018) Hypocalcemia and bone mineral density changes following denosumab treatment in end-stage renal disease patients: a meta-analysis of observational studies. Osteoporos Int 29:1737–1745
15. Dave V, Chiang CY, Booth J et al (2015) Hypocalcemia post denosumab in patients with chronic kidney disease stage 4–5. Am J Nephrol 41:129–137
16. Gnant M, Pfeiler G, Dubsky PC et al (2015) Adjuvant denosumab in breast cancer (ABCSG-18): a multicentre, randomised, double-blind, placebo-controlled trial. Lancet 386:433–443
17. Rachner TD, Göbel A, Jaschke NP et al (2020) Challenges in preventing bone loss induced by aromatase inhibitors. J Clin Endocrinol Metab 105:dgaa463. https://doi.org/10.1210/clinem/dgaa463
18. Lipton A (2009) Randomized trial of denosumab in patients receiving adjuvant aromatase inhibitors for nonmetastatic breast cancer. Breast Dis 2:195–196
19. The Lancet Diabetes Endocrinology (2021) Osteoporosis: overlooked in men for too long. Lancet Diabetes Endocrinol 9:1. https://doi.org/10.1016/S2213-8587(20)30408-3
20. Langdahl BL, Teglbjærg CS, Ho PR et al (2015) A 24-month study evaluating the efficacy and safety of denosumab for the treatment of men with low bone mineral density: results from the ADAMO trial. J Clin Endocrinol Metab 100:1335–1342
21. Nakamura T, Matsumoto T, Sugimoto T et al (2014) Clinical Trials Express: fracture risk reduction with denosumab in Japanese postmenopausal women and men with osteoporosis: denosumab fracture intervention randomized placebo controlled trial (DIRECT). J Clin Endocrinol Metab 99:2599–2607
22. Brown JE, Handforth C, Compston JE et al (2020) Guidance for the assessment and management of prostate cancer treatment-induced bone loss. A consensus position statement from an expert group. J Bone Oncol 25:100311. https://doi.org/10.1016/j.jbo.2020.100311
23. Kanis JA, Harvey NC, Cooper C et al (2016) A systematic review of intervention thresholds based on FRAX: a report prepared for the National Osteoporosis Guideline Group and the International Osteoporosis Foundation. Arch Osteoporos 11:25. https://doi.org/10.1007/s11657-016-0278-z

24. Serpa Neto A, Tobias-Machado M, Esteves MA et al (2012) Bisphosphonate therapy in patients under androgen deprivation therapy for prostate cancer: a systematic review and meta-analysis. Prostate Cancer Prostatic Dis 15:36–44

25. Smith MR, Egerdie B, Hernández Toriz N et al (2009) Denosumab in men receiving androgen-deprivation therapy for prostate cancer. N Engl J Med 361:745–755

26. Doria C, Leali PT, Solla F et al (2016) Denosumab is really effective in the treatment of osteoporosis secondary to hypogonadism in prostate carcinoma patients? A prospective randomized multicenter international study. Clin Cases Miner Bone Metab 13:195–199

27. Fizazi K, Carducci M, Smith M et al (2011) Denosumab versus zoledronic acid for treatment of bone metastases in men with castration-resistant prostate cancer: a randomised, double-blind study. Lancet 377:813–822

28. Smith MR, Saad F, Coleman R et al (2012) Denosumab and bone-metastasis-free survival in men with castration-resistant prostate cancer: results of a phase 3, randomised, placebo-controlled trial. Lancet 379:39–46

29. Fardet L, Petersen I, Nazareth I (2015) Monitoring of patients on long-term glucocorticoid therapy: a population-based cohort study. Medicine 94:e647. https://doi.org/10.1097/MD.000 0000000000647

30. Buckley L, Guyatt G, Fink HA et al (2017) 2017 American College of Rheumatology guideline for the prevention and treatment of glucocorticoid-induced osteoporosis. Arthritis Rheumatol 69:1521–1537

31. Compston J, Cooper A, Cooper C et al (2017) UK clinical guideline for the prevention and treatment of osteoporosis. Arch Osteoporos 12:43. https://doi.org/10.1007/s11657-017-0324-5

32. Saag KG, Emkey R, Schnitzer TJ et al (1998) Alendronate for the prevention and treatment of glucocorticoid-induced osteoporosis. Glucocorticoid-induced osteoporosis intervention study group. N Engl J Med 339:292–299

33. Reid DM, Hughes RA, Laan RF et al (2000) Efficacy and safety of daily risedronate in the treatment of corticosteroid-induced osteoporosis in men and women: a randomized trial. European corticosteroid-induced osteoporosis treatment study. J Bone Miner Res 15:1006–1013

34. Reid DM, Devogelaer JP, Saag K et al (2009) Zoledronic acid and risedronate in the prevention and treatment of glucocorticoid-induced osteoporosis (HORIZON): a multicentre, double-blind, double-dummy, randomised controlled trial. Lancet 373:1253–1263

35. Recker RR, Gallagher R, MacCosbe PE (2005) Effect of dosing frequency on bisphosphonate medication adherence in a large longitudinal cohort of women. Mayo Clin Proc 80:856–861

36. Saag KG, Shane E, Boonen S et al (2007) Teriparatide or alendronate in glucocorticoid-induced osteoporosis. N Engl J Med 357:2028–2039

37. Saag KG, Pannacciulli N, Geusens P et al (2019) Denosumab versus risedronate in glucocorticoid-induced osteoporosis: final results of a twenty-four-month randomized, double-blind, double-dummy trial. Arthritis Rheumatol 71:1174–1184

38. Buckley L, Guyatt G, Fink HA et al (2017) American College of Rheumatology Guideline for the prevention and treatment of glucocorticoid-induced osteoporosis. Arthritis Care Res 69:1095–1110

39. American College of Rheumatology (2022) American College of Rheumatology guideline for the prevention and treatment of glucocorticoid-induced osteoporosis. https://www.rheumatol ogy.org/Practice-Quality/Clinical-Support/Clinical-Practice-Guidelines/Glucocorticoid-Ind uced-Osteoporosis. Accessed 02 Apr 2023

40. Mok CC, Ho LY, Ma KM (2015) Switching of oral bisphosphonates to denosumab in chronic glucocorticoid users: a 12-month randomized controlled trial. Bone 75:222–228

41. Iseri K, Iyoda M, Watanabe M et al (2018) The effects of denosumab and alendronate on glucocorticoid-induced osteoporosis in patients with glomerular disease: a randomized, controlled trial. PLoS ONE 13:e0193846. https://doi.org/10.1371/journal.pone.0193846

42. Saag KG, Wagman RB, Geusens P et al (2018) Denosumab versus risedronate in glucocorticoid-induced osteoporosis: a multicentre, randomised, double-blind, active-controlled, double-dummy, non-inferiority study. Lancet Diabetes Endocrinol 6:445–454

43. Geusens P, Bevers MS, van Rietbergen B et al (2022) Effect of denosumab compared with risedronate on bone strength in patients initiating or continuing glucocorticoid treatment. J Bone Miner Res 37:1136–1146

44. Matsuno H (2016) Assessment of distal radius bone mineral density in osteoporosis patients receiving denosumab, including those with rheumatoid arthritis and those receiving oral glucocorticoids. Drugs R D 16:347–353

45. Hu MI, Glezerman IG, Leboulleux S et al (2014) Denosumab for treatment of hypercalcemia of malignancy. J Clin Endocrinol Metab 99:3144–3152

46. Thosani S, Hu MI (2015) Denosumab: a new agent in the management of hypercalcemia of malignancy. Future Oncol 11:2865–2871

47. Stewart AF (2005) Clinical practice. Hypercalcemia associated with cancer. N Engl J Med 352:373–379

48. Major P, Lortholary A, Hon J et al (2001) Zoledronic acid is superior to pamidronate in the treatment of hypercalcemia of malignancy: a pooled analysis of two randomized, controlled clinical trials. J Clin Oncol 19:558–567

49. Major PP, Coleman RE (2001) Zoledronic acid in the treatment of hypercalcemia of malignancy: results of the international clinical development program. Semin Oncol 28:17–24

50. Henry DH, Costa L, Goldwasser F et al (2011) Randomized, double-blind study of denosumab versus zoledronic acid in the treatment of bone metastases in patients with advanced cancer (excluding breast and prostate cancer) or multiple myeloma. J Clin Oncol 29:1125–1132

51. Raje N, Terpos E, Willenbacher W et al (2018) Denosumab versus zoledronic acid in bone disease treatment of newly diagnosed multiple myeloma: an international, double-blind, double-dummy, randomised, controlled, phase 3 study. Lancet Oncol 19:370–381

52. Diel IJ, Body JJ, Stopeck AT et al (2015) The role of denosumab in the prevention of hypercalcaemia of malignancy in cancer patients with metastatic bone disease. Eur J Cancer 51:1467–1475

53. Raje N, Roodman GD, Willenbacher W et al (2018) A cost-effectiveness analysis of denosumab for the prevention of skeletal-related events in patients with multiple myeloma in the United States of America. J Med Econ 21:525–536

54. Huang SY, Yoon SS, Shimizu K et al (2020) Denosumab versus zoledronic acid in bone disease treatment of newly diagnosed multiple myeloma: an international, double-blind, randomized controlled phase 3 study-Asian subgroup analysis. Adv Ther 37:3404–3416

55. Rosen LS, Gordon D, Tchekmedyian NS et al (2004) Long-term efficacy and safety of zoledronic acid in the treatment of skeletal metastases in patients with nonsmall cell lung carcinoma and other solid tumors: a randomized, Phase III, double-blind, placebo-controlled trial. Cancer 100:2613–2621

Romosozumab: Clinical Applications, Outcomes, and Perspectives in Osteoporosis Treatment

Giacomina Brunetti

Abstract Osteoporosis is a major worldwide health problem affecting bone heath. One of the most recent new therapeutic strategies that has been introduced is romosozumab. It is a humanized antibody directed against sclerostin. This is an antagonist of the Wnt signaling pathway, which is known to affect the activity of osteoblasts directly and osteoclast function indirectly. The crucial role of this molecule has been further sustained by clinical results associated with different bone diseases, including osteoporosis. Initially, results on animal models encouraged the introduction of romosozumab for osteoporosis treatment, since then different clinical trials have been organized including ARCH, FRAME, STRUCTURE, an extended trial and BRIDGE. Very interesting results have emerged thus favoring the approval of its use for osteoporosis management by the FDA and EMA, with the only recommendation being to pay particular attention to patients who have previously had a myocardial infarct.

Keywords Osteoporosis · Romosozumab · Wnt pathway · Sclerostin

1 Introduction

Osteoporosis is a major worldwide health problem affecting bone heath [1]. In this disease, bones lose their mechanical strength as well as mineral density and thus architecture, consequently they appear as a "porous bone" [2]. The main clinical feature of osteoporosis is represented by fractures. In particular, it has been reported that over the age of 50 one in three women and one in five men will present an osteoporotic fracture [2, 3]. The most affected sites are hip and spine, and these can lead to the development of severe pain, disability and even death. An estimation of the osteoporosis burden in Europe has been reported in which 25.5 million women versus 6.5 million men were counted (a ratio of 4:1) together with 4.3 million

G. Brunetti (✉)
Department of Biosciences, Biotechnologies and Environment, University of Bari "Aldo Moro", Bari, Italy
e-mail: giacomina.brunetti@uniba.it

© The Author(s), under exclusive license to Springer Nature Singapore Pte Ltd. 2023
A. H. Choi and S. Yik Lim (eds.), *Pharmacological Interventions for Osteoporosis,*
Tissue Repair and Reconstruction, https://doi.org/10.1007/978-981-99-5826-9_3

fragility fractures [2, 3]. These numbers are also associated with fracture-related deaths that are comparable or exceed some of the most widespread causes of death (cancer, diabetes, cardiovascular disease). Between 2019 and 2034 subjects over 50 years old are estimated to increase by 11.4% in men and women and in parallel the annual number of osteoporotic fractures will increase by about 25% [2, 3]. Thus, it is important to counteract this disease in order to improve the quality of life of people worldwide, also considering the increase in longevity. Very recently, new pharmacological approaches have been proposed and the latest new entry is represented by romosozumab, a humanized IgG2 monoclonal antibody able to interact with and inhibit sclerostin. To understand the role of this molecule it is important to know what happens to bone that leads to osteoporosis [4]. It develops as an imbalance between bone formation by osteoblasts and bone resorption by osteoclasts, two tightly regulated processes that characterize lifelong bone remodeling [5]. In fact, it is required for bone growth and adaptation to loading variations as well as repair. One of the key pathways involved in bone remodeling is represented by the WNT signaling pathway.

1.1 Canonical WNT Signaling Pathway and Sclerostin

The Wnt pathway is the name of the numerous signaling cascades that can be triggered by 19 different Wnt glycoproteins [summarized in 4]. The most well-known is the canonical pathway which involves the translocation of β-catenin into the nucleus with the consequent transcription of genes affecting osteoblast and osteoclast activity. To trigger a signal, Wnt ligands interact with the Frizzled (FZD) receptor and Low-density lipoprotein receptor-related protein 5/6 (LRP5/6) co-receptors which determine the phosphorylation of the LRP5/6 cytoplasmic tail together with the consequent recruitment of the destruction complex of β-catenin thus resulting in the inhibition of its degradation, and thus β-catenin translocation into the nucleus.

Wnt signaling is controlled by different endogenous molecules that can act by either binding Wnt glycoproteins directly or the LRP5/6 coreceptors. The Frizzled related proteins (sFRPs) and Wnt inhibitory factor (WIF-1) are direct binders of Wnt, whereas Dickkopf (DKK) and sclerostin (the product of SOST gene) interact with LRP5/6. Due to the key role of Wnt signaling directly in the regulation of osteoblast differentiation, and indirectly in that of osteoclasts, different studies have been developed to discover the molecules regulating this pathway, also because in the meantime the involvement of the dysregulation of this pathway in bone diseases has been demonstrated by different authors [6]. It is important to detail that DKK1 is ubiquitously distributed in different tissues, whereas sclerostin showed a unique distribution in bone, although recently its secretion by skeletal muscle has been reported [7]. Sclerostin has been associated with the reduction of bone mineral density (BMD) at different sites, patients and mice with homozygous mutation in SOST displayed sclerostosis, characterized by the strong increase in bone mass and density, as well as van Buchen disease [described in 4]. On the contrary, the overexpression of SOST in transgenic mice is associated with low bone mass. Mechanical loading strongly affects the expression of this molecule, in fact mechanical stimulation decreases SOST expression, whereas unloading is associated with the increase in SOST expression [8, 9]. Increased levels of sclerostin have also been found in osteoporotic post-menopausal woman, which leads to development of its neutralizing antibody Romosozumab.

2 Romosozumab Treatment of Osteoporosis in Postmenopausal Women

Results from four different randomized controlled trials have been reported: the Active-contRolled fraCture study in postmenopausal women with osteoporosis at High risk (ARCH; NCT01631214) [10], the FRActure study in postmenopausal women with osteoporosis (FRAME; NCT01575834) [11], the Study evaluating effect of RomosozUmab compared with teriparatide in postmenopaUsal women with osteoporosis at high risk for fracture previously treated with bisphosphonate therapy (STRUCTURE; NCT01796301) [12], and the romosozumab Phase 2 dose-finding study extension (Phase 2 extension; NCT00896532) [13–15].

The ARCH study [10] enrolled 4093 osteoporotic women (age 55–90 years old) with a previous fragility fracture to be treated for 12 months with a double-blinded subcutaneous 210 mg romosozumab once monthly or weekly oral 70 mg alendronate, followed by open-label weekly oral 70 mg alendronate in both groups. The key finding of this study is the reduction of vertebral fracture incidence by 48% after 2 years of treatment in the romosozumab-alendronate group, with respect to the alendronate-alendronate group. A reduction in the incidence of clinical fractures of about 27% after 33 months of treatment in the romosozumab-alendronate group with respect to the alendronate-alendronate group is reported in the same study. Additional information includes the mean BMD increase from the baseline in the romosozumab group after 12 months of treatment in different bone segments: 13.7% in lumbar spine, 6.2% in total hip, and 4.9% at the femoral neck. Differently, at the same time point, the change in mean BMD from the baseline comparing the romosozumab group with the alendronate group were 8.7% in lumbar spine, 3.3% in total hip and 3.2% at the femoral neck. After 24 months of treatment the variation in the mean BMD from the baseline in the romosozumab group were 15.2% in lumbar spine, 7.1% in total hip and 5.9% at the femoral neck. Different results can be observed by evaluating the mean BMD change from the baseline by comparing the romosozumab group with the placebo group: 8.0% in lumbar spine, 3.7% in total hip and 3.8% at the femoral neck. Furthermore, following 33 months of treatment (the time of primary analysis) a significant reduction in non-vertebral fractures and hip fractures was reported.

The FRAME study [11] included 7180 women (55–90 years old), with a T-score of −2.5 to −3.5 at the femoral neck or in total hip, who were treated monthly for

12 months with double-blinded subcutaneous 210 mg romosozumab or a placebo, after which for a further 12 months both groups received open-label subcutaneous 60 mg denosumab every 6 months. The key findings arising from this study are the following: in the romosozumab-denosumab group compared with the placebo-denosumab group after 12 months of treatment the incidence of vertebral fracture was reduced by about 73%, whereas after 24 months the incidence of vertebral fractures is diminished by about 75%. However, additional key points were reported. In detail, after 1 year of treatment the change in BMD from the baseline in the romosozumab group were 13.3% in lumbar spine, 6.8% in total hip, and 5.2% at the femoral neck. Differently, after 1 year of treatment the BMD changes from the baseline, comparing the romosozumab group to placebo groups were 13.3% in lumbar spine, 6.9% in total hip and 5.9% at the femoral neck. After 2 years of treatment, more pronounced effects were reported. In detail, the changes in mean BMD from the baseline in the romosozumab group were 17.6% in lumbar spine, 8.8% in total hip and 6.6% at the femoral neck. The following changes in BMD were found comparing the romosozumab-denosumab group with the placebo-denosumab group: 12.6% in lumbar spine, 6.0% in total hip and 6.0% at the femoral neck. There were important results in this trial after 36 months of treatment. In detail, the changes in mean BMD from the baseline in the romosozumab group were 18.1% in lumbar spine, 9.4% in total hip and 8.2% at the femoral neck. The following changes in BMD were found comparing the romosozumab-denosumab group to the placebo-denosumab group: 10.5% in lumbar spine, 5.2% in total hip and 4.8% at the femoral neck.

The STRUCTURE trial [12] had 436 women (55–90 years old), with a history of nonvertebral fracture over 50 years old or a vertebral fracture and T-score of ≤−2.5 in the lumbar spine, total hip, or at the femoral neck who had been orally treated with bisphosphonate (BP) therapy for ≥3 years and oral weekly alendronate for ≥1 year prior to screening. Women (218) were randomly designated to receive monthly open-label subcutaneous 210 mg romosozumab or the other 218 women received daily subcutaneous 20 μg teriparatide for 1 year. The key finding of this study reported the mean percentage change from the baseline in the total hip areal BMD as 2.6% in the romosozumab group and −0.6% in the teriparatide group. After 6 months of treatment in this trial the changes in mean BMD from the baseline in the romosozumab group were already 7.2% in lumbar spine, 2.3% in total hip and 2.1% at the femoral neck; in contrast to the teriparatide group the BMD changes were 3.5% in lumbar spine, −0.8% in total hip and −1.1% at the femoral neck, a different hip strength was reported for the two groups, 2.1% and −1.0%, respectively. After 1 year of treatment interesting results were reported: the changes in mean BMD from the baseline in the romosozumab group were 9.8% in lumbar spine, 2.9% in total hip and 3.2% at the femoral neck. Differently, for the romosozumab group after 1 year of treatment for the teriparatide group the results were 5.4% in lumbar spine, −0.5% in total hip and −0.2% at the femoral neck, at this time the hip strength for the two groups was 2.5% and −0.7%, respectively. The romosozumab Phase 2 dose-finding study and its extensions [13–15] had 419 women enrolled (55–85 years old) with a T-score of ≤−2.0 and ≥−3.5 in the lumbar spine, total hip, or at the femoral neck into multiple arms and interventions over a 6-year period. The key findings of this study showed that all doses of romosozumab increased lumbar spine BMD. However, the monthly subcutaneous dose of 210 mg of romosozumab determined the greatest gain in BMD in lumbar spine, corresponding to an increase of 11.3% compared to the placebo (0.1% reduction), teriparatide increase (7.1%) and alendronate increase (4.1%). The extension study reports data in the following 48–72 months. In patients, not receiving further treatment the BMD changes were -10.8% in lumbar spine, -6.4% in total hip and -5.9% at the femoral neck. Differently, for patients receiving one dose of zoledronic acid after one year of romosozumab the BMD changes were the following: -0.8% in lumbar spine, 0.1% in total hip and 0.5% at the femoral neck.

Recently, from a prospective, observational, and multicenter study 63 treatment-naïve osteoporotic patients (mean age 72.6 years) were enrolled to be treated with romosozumab for 12 months in order to investigate the best biomarker able to predict the augmented BMD in the hip. The authors found that the baseline serum levels of the bone formation marker PINP were the only molecule able to differentiate the responders from non-responders in the BMD increase [16].

2.1 Histomorphometrical Studies on Romosozumab Treatment

Romosozumab treatment in phase II and III trials has been linked with the increase in bone formation markers in the first two months of treatment, but they returned to the baseline levels after 1 year of treatment. Conversely, the markers of bone resorption reduced in 1 year of treatment [11, 13]. Histomorphometric studies have also been performed on bone biopsies obtained from a FRAME subgroup following 2 and 12 months of treatment [17]. The obtained results were in line with the variations in the described bone markers. Similar results were previously obtained in animal studies, but in these models the evaluation of active bone forming surface in trabecular and endocortical bone together with bone modeling and remodeling lead to the demonstration that romosozumab affects bone formation mainly by modeling [18, 19]. Studies in rat showed that a modeling-based bone formation (MBBF) following sclerostin inhibition started through lining cell activation, continued with osteoprogenitor cells proliferation thus sustaining prolonged activity of the modeling-based formation unit [20–22]. These studies have been extended to biopsies obtained from the FRAME study, but only in 14 placebo and 15 romosozumab group patients after 2 months treatment [23]. In detail, Eriksen et al. report that the MBBF percentage with respect to the total bone surface was augmented in the romosozumab group with respect to the placebo group on the trabecular bone (18.0% vs. 3.8%) as well as on endocortical bone (36.7% vs. 3%) [23]. Interestingly, the same authors report that this percentage was not significantly changed on periosteal bone [23]. The evaluation of remodeling-based bone formation (RBBF) did not show significant variation in the two groups. Thus, these results on human biopsies confirm the data obtained from animal studies [23].

Another study reported the evaluation of bone formation and resorption on bone biopsies obtained after 2 and 12 months of treatment arising from the FRAME study, demonstrating that bone formation evaluated as osteoid volume, increased at 2 months of treatment, but following 12 months of treatment the osteoid volume was reduced compared with the placebo [17]. This latter decrease was linked to an expansion of the formation period, a mineralization onset delay, together with a decrease in the mineral apposition rate compared with the placebo [17]. These effects on bone formation were parallel to the prolonged reduction of bone resorption. The inhibitory effect on resorption could contribute to the uncoupling of bone remodeling with consequent effects also on formation.

2.2 Importance of the Treatment Sequence Using Romosozumab

To assess the effect of treatment sequence on romosozumab anabolic role, Cosman et al. [24] evaluated the data arising from studies where romosozumab was injected before an antiresorptive drug (FRAME and ARCH) and compared them with results from trials where romosozumab was injected after an antiresorptive agent (STRUCTURE, phase 2 extension). Very interestingly, the authors found that when romosozumab was injected before an antiresorptive treatment, 12 months of romosozumab therapy determined a BMD increase in the total hip of about 6.0–6.2% together with a BMD increase in lumbar spine of 13.1–13.7%. Differently, when the 12 month-treatment of romosozumab was subsequent to alendronate administration, the BMD increased 2.9% in the total hip and 9.8% in the lumbar spine. Moreover, when romosozumab treatment was subsequent to denosumab, the BMD increased by 0.9% in total hip and 5.3% in lumbar spine BMD. Thus, considering these treatments lasting over two years, the most important effects on BMD are evident when the treatment begins with romosozumab and goes on with antiresorptive therapy. Importantly, more significant BMD enhancement determined a more efficient decrease in fracture risk, thus romosozumab treatment followed by the antiresorptive therapy may lead to a large reduction in fracture risk reduction efficacy.

3   Romosozumab Treatment of Osteoporosis in Men

An additional important study evaluated the efficacy of romosozumab treatment in osteoporotic men (55–90 years old): the placebo-controlled study evaluating the efficacy and safety of romosozumab in treating men with osteoporosis (BRIDGE) [25]. A total of 245 men with osteoporosis (T-score ≤ -2.5 or -1.5 prior fragility vertebral or non-vertebral fracture were enrolled. 210 mg Romosozumab or a placebo were administered subcutaneously monthly for 12 months. As key findings, a percentage change from the baseline of BMD in lumbar spine of 12.1% was recorded in the romosozumab group compared with 1.2% in the placebo group. Additional findings include a mean BMD change from the baseline after 6 months of treatment in the romosozumab group of 9.0% in the lumbar spine, 1.6% in the total hip and 1.2% at the femoral neck. Differently, following 12 months of treatment there was a mean BMD change from the baseline in the romosozumab group of 12.1% in the lumbar spine, 2.5% in the total hip and 2.2% at the femoral neck. Thus, the reported increase in the lumbar spine is similar to that observed in osteoporotic women enrolled for the FRAME and ARCH studies.

4 Romosozumab Treatment of Osteoporotic Patients with Chronic Kidney Disease and End-Stage Renal Disease

The ARCH and FRAME trials included patients with chronic kidney disease. In patients with a reduced estimated glomerular filtration rate (eGFR), a tendency towards a reduced BMD increase was observed, although a BMD increase was evident in the patients compared with the controls. Concerning fracture prevention, in FRAME and ARCH, at 12 months, the decrease in the incidence of new vertebral compression fractures was similar to that in patients with normal kidney activity, mild renal insufficiency, and moderate renal insufficiency respectively [26].

There are very few results on romosozumab treatment in the end-stage renal disease. In a study involving 96 Japanese patients at high risk for fracture on hemodialysis, subjects treated with romosozumab displayed significant BMD increases in the lumbar spine and at the femoral neck [27].

5 Safety

Adverse events were similar in subjects treated with romosozumab and controls in the FRAME and ARCH studies. Nasopharyngitis and back pain occurred in about 10% of subjects. Site reactions due to injection were frequently observed in patients receiving romosozumab.

Atypical femoral fractures as well as osteonecrosis of the jaw (ONJ) have also been registered for romosozumab treated patients. In detail, in the FRAME study two patients displayed ONJ compared with the placebo group. The first appeared after 1 year of treatment, the second was evident after 1 year of romosozumab followed by one dose of denosumab. Atypical femoral fractures were only reported in 1 patient after three months of treatment with romosozumab [11].

In the ARCH study, ONJ occurred in 1 subject who was treated with romosozumab followed by alendronate. In this latter condition 2 cases of atypical femoral fractures were also recorded [10]. In parallel, 4 cases of atypical femoral fractures and 1 of ONJ occurred in patients treated with alendronate for 2 years in the same study.

The very important factor that slowed down the use of romosozumab is related adverse cardiovascular (CVD) effects. However, its use has been avoided in patients registered with an infarction in the past year. No important differences were recorded in the FRAME study that involves a greater number of subjects (7180) with CVD. Differently, in the ARCH study (4093 subjects) and BRIDGE (245 subjects) more CVD events were reported, with a greater frequency of cardiac ischemic events and cerebrovascular events in subjects receiving romosozumab [10, 11, 25]. CVD events not associated with revascularization were numerically reduced in the group receiving romosozumab and alendronate.

The BRIDGE trial reported CVD in 8 subjects treated with romosozumab versus 2 receiving the placebo [25].

During its approval for use the Food and Drug Administration (FDA) evaluated all these CVD events without finding a direct relationship. Moreover, a meta-analysis of the FRAME, ARCH and BRIDGE trials failed to show a statistically significant increase in major adverse cardiac events (MACE), representing a vital endpoint used in clinical trials, which includes cardiovascular death, nonfatal myocardial infarction or nonfatal stroke) [28].

Moreover, the European Medicine Agency (EMA) expressed its position on CVD and romosozumab, stating that MACE risk cannot be excluded or confirmed, that the absolute risk difference is small and that a more detailed analysis could support the idea that MACE risk in the ARCH study is over-estimated [29].

Interestingly, Miller et al. analyze the role of CVD in romosozumab-treated patients considering Kidney function [26]. They reported that in the FRAME study, the romosozumab and placebo groups showed identical CVD events considering the same stage of renal functionality. Differently, in the ARCH study CVD occurred to a greater extent in subjects with normal renal activity. To explain this issue, it has been proposed that alendronate displays a greater cardioprotective effect than romosozumab [30]: this matter represents a fundamental discussion topic. It has been found in an observational study that BPs reduced CVD events by 28–75%. Furthermore, denosumab treatment also seems to be associated with a 46% increase in CVD events compared to that with BPs [31].

Thus, the question arises as to whether sclerostin could impact on CVD risk. Sclerostin is expressed by smooth muscle cells in the aorta, but different in vivo and in vitro studies have failed to find adverse action of sclerostin inhibition on cardio-vascular health [32]. Consistently, human patients with sclerostin deficiency did not display CVD [33]. Furthermore, sclerostin shows a protective effect in a murine model of atherosclerosis [34]. Consistently, low levels of circulating sclerostin have been associated to an increased risk of developing atherosclerosis [35]. All these findings led the FDA and EMA to recommend caution for romosozumab use.

However, a very recent systematic review evaluating CVD events in osteoporosis treatment, also including romosozumab trials, reports that the number of CVD events associate with osteoporosis management (Abaloparatide, Alendronate, denosumab, Hormonal, Romosozumab, teriparatide, zoledronate) is not statistically significant [36].

Considering all the evidence available, further surveillance is recommended.

6   Cost Considerations

The economic burden of healthcare also plays a crucial part in establishing the treatment of choice for osteoporosis management. In the USA economic analyses do not encourage romosozumab use for osteoporosis management [37], because it is a branded medication and it is more expensive than generic antiresorptive medications. Analogous considerations have emerged in the UK [37]. Due to its high costs in Sweden, it is recommended in patients with a very high-risk of fracture in postmenopausal women. In conclusion, treatment with Romosozumab transitioned to bisphosphates represents a good deal in terms of both osteoporosis treatment and economic burden [38, 39].

7   Conclusion

Romosozumab effect on osteoporosis management is very encouraging, in the future could became the preferential drug, however it is fundamental to remember to take particular care and attention to patients with previous stories of myocardial infarction and/or other CVD. Thus, surveillance need!

References

1. Qaseem A, Hicks LA, Etxeandia-Ikobaltzeta I et al (2023) Pharmological treatment of primary osteoporosis or low bone mass to prevent fractures in adults: a living clinical guideline from the American college of physicians. Ann Intern Med. https://doi.org/10.7326/M22-1034
2. Willers C, Norton N, Harvey NC et al (2022) Osteoporosis in Europe: a compendium of country-specific reports. Arch Osteopor 17:23
3. Kanis JA, Norton N, Harvey NC et al (2021) SCOPE 2021: a new scorecard for osteoporosis in Europe. Arch Osteop 16:82
4. Gao Y, Chen N, Fu Z, Zhang Q (2023) Progress of Wnt signaling pathway in osteoporosis. Biomolecules 13:483
5. Mirza F, Canalis E (2015) Secondary osteoporosis: pathophysiology ana management. Eur J Endo 173:R131–R151
6. Dincel AS, Jørgensen NR, IOF-IFCC Joint Committee on Bone Metabolism (C-BM) (2023) New emerging biomarkers for bone disease: sclerostin and Dickkopf-1 (DKK1). Calcif Tissue Int 112:243–257
7. Magarò MS, Bertacchini J, Florio F et al (2021) Identification of sclerostin as a putative new myokine involved in the muscle-to-bone crosstalk. Biomedicines 9:71

8. Lin C, Jiang X, Dai Z et al (2009) Sclerostin mediates bone response to mechanical unloading through antagonizing Wnt/beta-catenin signaling. J Bone Miner Res 24:1651–1661

9. Spatz JM, Wein MN, Gooi JH et al (2015) The Wnt inhibitor sclerostin is up-regulated by mechanical unloading in osteocytes in vitro. J Biol Chem 290:16744–16758

10. Saag KG, PetersenJ BML et al (2017) Romosozumab or alendronate for fracture prevention in women with osteoporosis. N Engl J Med 377:1417–1427

11. Cosman F, Crittenden DB, Adachi JB et al (2016) Romosozumab treatment in postmenopausal women with osteoporosis. N Engl J Med 375:1532–1543

12. Langdahl BL, Libanati C, Crittenden DB et al (2017) Romosozumab (sclerostin monoclonal antibody) versus teriparatide in postmenopausal women with osteoporosis transitioning from oral bisphosphonate therapy: a randomized, open-label, Phase 3 trial. Lancet 390:1585–1594

13. McClung MR, Grauer A, Boonen S et al (2014) Romosozumab in postmenopausal women with low bone mineral density. N Engl J Med 370:412–420

14. Kendler DL, Bone HG, Massari F et al (2019) Bone mineral density gains with a second 12-month course of romosozumab therapy following placebo or denosumab. Osteopor Int 30:2437–2448

15. McLung MR, Bolognese MA, Brown JP et al (2021) Skeletal responses to romosozumab after 12 months of romosozumab. JBMR Plus 5:e10512

16. Kashii M, Kamatani T, Nagayama Y et al (2023) Baseline serum PINP level is associated with the increase in hip bone mineral density seen with Romosozumab treatment in previously untreated women with osteoporosis. Osteoporos Int 34:563–572

17. Chavassieux P, Chapurlat R, Portero-Muzy N et al (2019) Bone-forming and antiresorptive effects of romosozumab in postmenopausal women with osteoporosis: bone histomorphometry and microcomputed tomography analysis after 2 and 12 months of treatment. J Bone Miner Res 34:1597–1608

18. Boyce RW, Niu QT, Ominsky MS (2017) Kinetic reconstruction reveals time-dependent effects of romosozumab on bone formation and osteoblast function in vertebral cancellous and cortical bone in cynomolgus monkeys. Bone 101:77–87

19. Ominsky MS, Niu QT, Li C et al (2014) Tissue-level mechanisms responsible for the increase in bone formation and bone volume by sclerostin antibody. J Bone Miner Res 29:1424–1430

20. Boyce RW, Brown D, Felx M et al (2018) Decreased osteoprogenitor proliferation precedes attenuation of cancellous bone formation in ovariectomized rats treated with sclerostin antibody. Bone Rep 8:90–94

21. Greenbaum A, Chan KY, Dobreva T et al (2017) Bone CLARITY: clearing, imaging, and computational analysis of osteoprogenitors within intact bone marrow. Sci Transl Med 9:eaah6518

22. Kim SW, Lu Y, Williams EA et al (2017) Sclerostin antibody administration converts bone lining cells into active osteoblasts. J Bone Miner Res 32:892–901

23. Eriksen EF, Chapurlat R, Boyce RW et al (2022) Modeling-based bone formation after 2 months of romosozumab treatment: results from the FRAME clinical trial. J Bone Miner Res 37:36–40

24. Cosman F, Kendler DL, Langdahl BL et al (2022) Romosozumab and antiresorptive treatment: the importance of treatment sequence. Osteopor Int 33:1243–1256

25. Lewiecki EM, Blicharski T, Goemaere S et al (2018) A Phase III trial randomized placebo-controlled trial to evaluate efficacy and safety of romosozumab in men with osteoporosis. J Clin Endocrinol Metab 103:3183–3193

26. Miller PD, Adachi JD, Albergaria BH et al (2022) Efficacy and safety of romosozumab among postmenopausal women with osteoporosis and mild-to-moderate chronic kidney disease. J Bone Miner Res 37:1437–1445

27. Sato M, Inaba M, Yamada S et al (2021) Efficacy of romosozumab in patients with osteoporosis on maintenance hemodialysis in Japan; an observational study. J Bone Miner Metab 39:1082–1090

28. Bovijn J, Krebs K, Chen CY et al (2020) Evaluating the cardiovascular safety of sclerostin inhibition using evidence from meta-analysis of clinical trials and human genetics. Sci Transl Med 12:eaay6570

29. European Medicines Agency (2019) Evenity. https://www.ema.europa.eu/en/medicines/human/EPAR/evenity. Accessed 15 Feb 2023
30. Adami G, Gatti D, Fassio A et al (2023) Cardiovascular safety of romosozumab: new insights from postmenopausal women with chronic kidney disease. J Bone Miner Res 38:354–355
31. Seeto AH, Abrahamsen B, Ebeling PR et al (2021) Cardiovascular safety of denosumab across multiple indications: a systematic review and meta-analysis of randomized trials. J Bone Miner Res 36:24–40
32. Turk JR, Deaton AM, Yin J et al (2020) Nonclinical cardiovascular safety evaluation of romosozumab, an inhibitor of sclerostin for the treatment of osteoporosis in postmenopausal women at high risk of fracture. Reg Toxicol Pharmacol 115:104697. https://doi.org/10.1016/j.yrtph.2020.104697
33. van Lierop A, Appelman-Dijkstra NM, Papapoulos SE et al (2017) Sclerostin deficiency in humans. Bone 96:51–62
34. Krishna SM, Seto SW, Jose RJ et al (2017) Wnt signaling pathway inhibitor sclerostin inhibits angiotensin II-induced aortic aneurysm and atherosclerosis. Arterioscler Thromb Vasc Biol 37:553–556
35. Zheng J, Smith GD, Kavousi M et al (2022) Lowering of circulating sclerostin may increase risk of atherosclerosis and its risk factors: evidence from a genome-wide association meta-analysis followed by Mendelian randomization. MedRxiv. https://doi.org/10.1101/2022.06.13.22275915
36. Seeto AH, Tadrous M, Gebre AK et al (2023) Evidence for the cardiovascular effects of osteoporosis treatments in randomized trials of post-menopausal women: a systematic review and Bayesian network meta-analysis. Bone 167:116610
37. McClung MR (2021) Role of bone-forming agents in the management of osteoporosis. Aging Clin Exp Res 33:775–791
38. Soreskog E, Lindberg I, Kanis JA et al (2021) Cost-effectiveness of romosozumab for the treatment of postmenopausal women with severe osteoporosis at high risk of fracture in Sweden. Osteoporos Int 32:585–594
39. Goeree R, Burke N, Jobin M et al (2022) Cost-effectiveness of romosozumab for the treatment of postmenopausal women at very high risk of fracture in Canada. Arch Osteoporos 17:71

Parathyroid Hormone and Parathyroid Hormone Analogs: Clinical Applications and Perspectives in Osteoporosis Treatment

Sian Yik Lim

Abstract Osteoporosis is a skeletal disorder characterized by decreased bone strength, leading to increased fracture risk. In this article, we discuss the use of parathyroid hormone analogs (teriparatide and abaloparatide) in the treatment of osteoporosis. Parathyroid hormone analogs are potent osteoanabolics developed to treat osteoporosis. We discuss important topics pertinent to clinical care to provide clinicians with helpful information in their daily practice when prescribing parathyroid hormone analogs for osteoporosis treatment. We also discuss the potential side effects and strategies to mitigate these side effects.

Keywords Osteoporosis · Bisphosphonates · Skeletal disorder

1 Introduction

Osteoporosis is a skeletal disorder characterized by reduced bone strength, increasing fracture risk [1]. Teriparatide, one of the first osteoanabolic agents developed to treat osteoporosis, was approved by Federal Drug Administration (FDA) in 2002. Subsequently, abaloparatide was developed and approved by the FDA in 2017 to treat osteoporosis in postmenopausal women. The development of teriparatide and abaloparatide for osteoporosis stemmed from observations of anabolic effects on the bone with intermittent low-dose administration of parathyroid hormone (PTH) [2]. This is despite the catabolic effects on the skeleton when there is continuous exposure to the parathyroid hormone, as seen in conditions such as primary hyperparathyroidism. This article discusses parathyroid hormone and parathyroid hormone analogs in the treatment of osteoporosis.

S. Y. Lim (✉)
Hawaii Pacific Health Medical Group, Honolulu, HI, USA
e-mail: limsianyik@gmail.com

Bone and Joint Center, Pali Momi Medical Center, 98-1079 Moanalua Road, Suite 300, Aiea, HI 96701, USA

© The Author(s), under exclusive license to Springer Nature Singapore Pte Ltd. 2023   47
A. H. Choi and S. Yik Lim (eds.), *Pharmacological Interventions for Osteoporosis*,
Tissue Repair and Reconstruction, https://doi.org/10.1007/978-981-99-5826-9_4

2 Chemical Structure of Parathyroid Hormone, Abaloparatide, and Teriparatide

Parathyroid hormone cells produce PTH, an 84 amino acid peptide hormone. Teriparatide is a truncated form of PTH consisting of 1–34 amino acids of the PTH hormone. PTHRP was initially noted as the cause of hypercalcemia related to tumors. PTHRP has properties similar to PTH. However, it causes less hypercalcemia and has fewer resorptive effects on the skeleton. Abaloparatide is an analog of parathyroid hormone-related protein (PTHRP) consisting of the first 34 amino acids of human PTHRP [3]. Twenty amino acids are exchanged in abaloparatide yielding a 42% homology with human PTH (1–34) or teriparatide [3].

3 Parathyroid Hormone, Abaloparatide, and Teriparatide: Mechanism of Action, Pharmacokinetics

3.1 Distinctive PTH Effects on RANK and Wnt Signalling Pathways

PTH plays an essential role in regulatory pathways of bone formation and resorption by binding to the parathyroid hormone (PTH) receptor. On the one hand, PTH increases the RANKL/OPG ratio, activating osteoclasts and increasing bone resorption. On the other hand, PTH also down-regulates sclerostin, leading to activation of the wnt signaling pathway and bone formation. Whether or not the effect is anabolic or catabolic depends on the time interval of PTH exposure. It is thought that continuous administration of PTH leads to more pronounced catabolic effects via its effect on the RANK signaling pathway. PTH administered intermittently at low doses leads to a predominantly anabolic effect. Teriparatide and abaloparatide are parathyroid hormone one receptor agonists (PTH1R). Abaloparatide binds more selectively to RG conformation of PTH1R over the R0 confirmation than teriparatide. Because the RG conformation is associated with a more transient binding of the ligand, it is hypothesized that abaloparatide produces more transient signaling responses leading to a higher anabolic effect than teriparatide [4].

3.2 PTH and PTH Analogs as Proremodeling Anabolic Agents

The skeletons' response to teriparatide, abaloparatide, and intermittent administration of PTH is an initial higher increase in bone formation without bone resorption (bone modeling). Bone modeling activation leads to an anabolic window where the bone is maximally formed [5]. Subsequently, bone remodeling becomes the predominant effect of PTH stimulation, with increased bone resorption and bone formation [6], with associated overflow modeling-based formation [7].

3.3   Pharmacokinetics and Pharmacodynamics

Teriparatide and abaloparatide undergo non-specific proteolytic degradation followed by renal elimination. Administration of teriparatide and abaloparatide leads to an increase in serum markers of bone formation: procollagen type 1 N-terminal propeptide (P1NP) and serum markers of bone resorption: C-terminal cross-linking telopeptide of type 1 collagen (CTX) [8, 9]. In the ACTIVE trial, P1NP increased to 93% above baseline in 1 month, and then declined to 45% above baseline by 18 months. CTX increased to 43% above baseline in 3 months, and then reduced to 20% above baseline at the end of treatment [4]. The rapid increase in P1NP, followed by a less rapid increase in CTX, is consistent with the concept of an anabolic window, where there is an initial increase in bone formation followed by a slower increase in bone resorption. While similar changes in P1NP and CTX were noted with teriparatide, gains were higher with teriparatide. The rise in CTX is less with abaloparatide (lower peak with a more gradual increase) than teriparatide (while an increase in P1NP in the first month is similar). Therefore, abaloparatide has a higher anabolic effect than teriparatide [4].

4   Efficacy

Clinical trials established the efficacy of teriparatide [10] and abaloparatide [11] (Table 1).

4.1   Teriparatide

Teriparatide effectively reduces vertebral fractures, non-vertebral fragility fractures, and hip fractures. The efficacy of teriparatide was demonstrated and established by the Fracture Prevention Trial [10]. The trial was a randomized, blinded, placebo-controlled study to assess teriparatide's efficacy in treating osteoporosis.

The trial included 1637 postmenopausal women with osteoporosis (mean age 69 years) from 99 centers in 17 countries. The study participants were ambulatory and were at least five years since menopause. Patients had a ≥1 moderate vertebral fracture or ≥2 mild vertebral fractures on thoracic or lumbar X-rays. Exclusion criteria included impaired hepatic function, creating of ≥2 mg/dL, and alcohol or

Table 1 Efficacy studies of teriparatide and abaloparatide

Study ID/study name/ trial design	Trial design/study context	Study population	Comparator groups/ denosumab dose	Primary endpoint	Conclusions
Teriparatide					
The fracture prevention trial [10]	Randomized control trial	1637 postmenopausal women with severe osteoporosis ≥1 moderate vertebral fracture, or ≥2 mild vertebral fracture	Subcutaneous teriparatide daily for 3 years or placebo **Teriparatide doses**: 20, 40 μg	(1) New vertebral fracture (X-ray) at 3 years The study stopped early due to preclinical studies showing increased risk of osteosarcoma in rats Median follow-up time 21 months	(1) New vertebral fracture reduced by 65–69% compared to placebo (2) Non-vertebral fracture reduced by 35–40% compared to placebo (3) 20, 40 μg no difference in clinical efficacy (4) BMD gain compared to placebo: 20 μg LS: 9%, FN 3%, 40 μg 20 μg LS: 13%, FN 6%

(continued)

Table 1 (continued)

Study ID/study name/ trial design	Trial design/study context	Study population	Comparator groups/ denosumab dose	Primary endpoint	Conclusions
Teriparatide or alendronat in glucocorticoid-induced osteoporosis [13, 14]	Randomized control trial	428 Patients ages 22–89 who received ≥5 mg/day of prednisone equivalent for ≥3 months preceding screening	Subcutaneous teriparatide daily for 3 years or alendronate 10 mg daily	(1) Change from baseline to last measurement of bone density at the lumbar spine	18 monthas (1) BMD gain compared to baseline: LS: Teriparatide 7.2%, Alendronate 3.4%, TH: Teriparatide 3.8%, Alendronate 2.4% (2) New vertebral fracture 0.6% in teriparatide group, 6.1% alendronate group ($p < 0.001$) (3) Non-vertebral fracture similar 36 monthas (1) BMD gain compared to baseline: LS: Teriparatide 11.0%, Alendronate 5.3%, TH: Teriparatide 5.2%, Alendronate 2.7% (2) New vertebral fracture 1.7% in teriparatide group, 2.7% alendronate group ($p < 0.001$) (3) Non-vertebral fracture similar

(continued)

Table 1 (continued)

Study ID/study name/ trial design	Trial design/study context	Study population	Comparator groups/ denosumab dose	Primary endpoint	Conclusions
Abaloparatide					
Abaloparatide comparator trial in vertebral endpoints (ACTIVE) [11]	Randomized control trial	2463 postmenopausal women	Subcutaneous teriparatide daily for 18 months or placebo or open-label teriparatide **Abaloparatide doses:** 80 μg	(1) New vertebral fracture (X-ray) at 6 months	(1) New vertebral fracture reduced by 86% compared to placebo (2) Non-vertebral fracture reduced by 43% compared to placebo Mean BMD change from Baseline, Placebo LS: 10.4%, TH: 4.3%, FN: 4.0%

drug abuse. Patients with urolithiasis within the past five years and diseases that affect calcium and bone metabolism were excluded.

Patients were randomized to receive a placebo (544 patients) or treatment: teriparatide 20 μg daily (541 patients) or teriparatide 40 μg daily (552 patients) injected subcutaneously daily. The primary outcome was a new onset vertebral compression fracture at three years. Other outcomes measured were new onset fractures and bone mineral density (BMD) changes.

The study was stopped early, at 20–24 months, due to preclinical osteosarcoma findings in rats in patients who received high doses of teriparatide. The average age of patients in the study was approximately 69 years, with well-matched baseline characteristics between both treatment arms. Patients had a mean T score of −2.6 and, on average, had more than two fractures per patient. Eighty one percent of patients had follow-up radiographic studies. The median follow-up was 21 months. The average duration of treatment was 18 months, with patients receiving teriparatide for up to 24 months.

Teriparatide 20 μg daily and teriparatide 40 μg daily led to a 65–69% relative rate reduction in new vertebral fractures. Teriparatide showed a 35–40% RRR of non-vertebral fragility fractures compared to the placebo. Compared to the placebo, teriparatide 20 μg daily and teriparatide 40 μg daily increased BMD in the lumbar spine by 9% and 13%, respectively. In the FN, the increase in BMD as compared to the placebo was 3% (teriparatide 20 μg daily) and 6% (teriparatide 40 μg daily) more. Although there was a more significant gain in BMD with 40 μg daily dosing, there was no significant difference regarding fracture prevention efficacy between 20 μg daily and 40 μg daily dosing. Hypercalcemia is more common with 40 μg daily dosing. Teriparatide's 20 μg daily dose is approved by the Food and Drug Administration (FDA) [8, 12].

4.2 Abaloparatide

Abaloparatide has been shown to reduce vertebral and non-vertebral fractures with an associated gain in bone mineral density. The efficacy of abaloparatide was demonstrated in the Abaloparatide Comparator Trial in Vertebral Endpoints (ACTIVE) trial [11]. ACTIVE was a phase 3 randomized, placebo-controlled study to assess the fracture reduction efficacy of abaloparatide 80 μg daily.

The ACTIVE trial included 2463 postmenopausal women with osteoporosis between ages 49 and 86. The study involved 28 study centers in 10 countries. Inclusion criteria included patients with prior fragility fracture: with ≥1 moderate vertebral fracture or ≥2 mild vertebral fractures, or a history of nonvertebral fracture (<5 years prior)-femoral neck, a spine T score of ≤−2.5 and age less than 65, femoral neck, a spine T score of ≤−2.0 if older than age 65. The study included patients with no history of a fragility fracture if the patient was >65 years of age and had a T score ≤−2.5 and >−5.0.

The study participants were randomized in a double-blinded fashion to receive a placebo, abaloparatide 80 μg subcutaneously daily, or open-label teriparatide 20 μg subcutaneously for 18 months. The primary outcome was a new onset vertebral compression fracture at six months. After 18 months, patients in the placebo and abaloparatide groups received alendronate for 24 months.

The average age of patients in the study was approximately 68.8 years, with well-matched baseline characteristics between both treatment arms. Patients had a mean T score of -2.1. About 24% of patients had a prevalent fracture, 31% reported a history of non-vertebral fracture in the past five years, and 37% had no prior fractures.

At month 6, the relative risk reduction of new morphometric vertebral fractures was 86% in patients receiving abaloparatide and 80% in patients receiving teriparatide compared to placebo. Over 18 months, the relative risk reduction of non-vertebral fractures was 43% with abaloparatide compared to placebo. Relative risk reduction for osteoporotic fractures was 70% with abaloparatide compared to placebo. Abaloparatide was associated with a gain in bone density in the lumbar spine, femoral neck, and total hip of 10.4%, 4.0%, and 4.3% compared to the placebo group. The gain in bone density in the hip occurred more rapidly than with teriparatide. Correlating with this, the Kaplan Meier curves for non-vertebral and osteoporotic fractures show an earlier separation of the abaloparatide curve compared to teriparatide versus placebo (approximately one year). This suggests early anti-fracture reduction efficacy with abaloparatide fulfilling a major unmet need for the patient with severe osteoporosis at higher risk of fracture (Table 2).

5   Safety and Tolerability

5.1   Teriparatide in Clinical Trials

In the Fracture Intervention Trial, teriparatide was well tolerated. There were no differences in number of deaths, hospitalizations, or cardiovascular disorders between the intervention and placebo groups. In the Fracture Intervention Trial for teriparatide, the most common side effects were occasional nausea and headache [10]. Other symptoms noted included dizziness, limb pain, and leg cramps. Due to the possibility of dizziness and orthostatic hypotension, patients should receive their injection while sitting [15].

5.2   Abaloparatide in Clinical Trials

In the ACTIVE trial, there were no differences regarding treatment-emergent adverse events, serious adverse events, or adverse events leading to death. Common side effects that led to discontinuation of abaloparatide were similar to teriparatide, including nausea, headache, and dizziness, in most cases mild-moderate in severity [11]. Due to the possibility of dizziness and orthostatic hypotension, patients should receive their injection while sitting [15].

5.3   Hypercalcemia

Subcutaneous injections of the parathyroid hormone have the most significant effect on calcium in the first 4–6 h after injection. In the Fracture Intervention Trial for teriparatide [10], when calcium was measured during that time interval, mild hypercalcemia 10.6 mg/dL occurred in 2% in the placebo group, 11% in the patients who received 20 μg teriparatide and 28% in patients who received 40 μg teriparatide. 95% of patients with hypercalcemia had a level less than 11.2 mg/dL in patients that received 20 μg teriparatide. The calcium returned to normal in about two-thirds of patients a few weeks later. In the ACTIVE trial [11], abaloparatide had lower rates of hypercalcemia than teriparatide (3.42% vs. 6.1%). A commonly used clinical protocol for monitoring teriparatide use is to check serum calcium one month and three months after starting teriparatide [16]. If no hypercalcemia is noted, the likelihood of hypercalcemia developing during the duration of treatment is lower. Ideally, calcium levels should be drawn 16 h or more after injection to avoid detecting clinically irrelevant transient calcium elevations [16]. With elevated calcium, the clinician can limit calcium intake to less than 1000 mg daily, especially if the patient takes calcium supplements. If hypercalcemia is persistent, consideration can be given to reducing teriparatide injections to alternate days [17].

5.4   Increased Risk of Osteosarcoma in Rats

When initially approved, teriparatide carried a boxed warning of osteosarcoma, limiting lifetime use to 24 months. The black box warning was based on data that rats receiving very high doses of teriparatide administered throughout the rat's lifespan were associated with the development of osteosarcoma [8, 18]. The most extended time period patients were on teriparatide was 24 months in the Fracture Intervention Trial. Due to the possibility of osteosarcoma in humans, the Fracture Prevention Trial studying the efficacy of teriparatide was terminated early [10]. The FDA approved teriparatide use with the abovementioned limitations and a black box warning. In a subsequent post-marketing surveillance study, no increased risk of osteosarcoma was noted in patients treated with teriparatide [19], which led to removing the box warning and limiting lifetime use.

6 Special Considerations

6.1 Chronic Kidney Disease

There is a lack of abaloparatide or teriparatide data in patients with chronic kidney disease. Post hoc analyses from the ACTIVE trial showed that there were no significant differences in anti-fracture efficacy and safety in regards to the use of these agents in patients with mild (eGFR 60 to <90 mL/min), moderate chronic kidney disease (<60 mL/min, with patients with eGF <37 mL/min, serum creatine >2.0 mg/dL excluded in the ACTIVE trial) [20].

6.2 Chronic Glucocorticoid Use

Glucocorticoids primarily affect osteoblastogenesis and osteoblast apoptosis and constitute a significant risk factor for osteoporosis. In patients on chronic glucocorticoids, teriparatide treatment led to a more substantial gain in the lumbar spine and hip bone density and a more significant reduction in vertebral compression fracture compared to patients who received alendronate at 18 and 36 months [13, 14]. The 36-month data showed the benefit of teriparatide treatment regarding BMD and vertebral fracture risk reduction was primarily in the first 24 months [13].

7 Long-Term Use of PTH and PTHrp Analogs

7.1 Teriparatide

However, FDA use of teriparatide was previously limited to 24 months [8]. This 24-month recommendation was based on the Fracture Prevention Trial, where the longest duration patient had been on teriparatide was 24 months. The 24-month limitation on the FDA label for teriparatide has been removed. However, no clear guidelines have been published on patients receiving teriparatide for more than 24 months. The FDA label has been modified to state, "Use of teriparatide for more than 24 months during the patient's lifetime should be considered only if a patient remains at or has returned to having a high risk for fracture". Various specialty organizations have defined a high risk for fracture (Table 2). Expert opinion suggests that teriparatide treatment for more than 24 months during the patient's lifetime may be reasonable in patients at very high/high risk for fracture: (1) unable to stop glucocorticoid treatment (2) multiple vertebral compression fractures at baseline but none on teriparatide (3) severe chronic obstructive pulmonary disease with vertebral compression fractures [16]. Other patients that may benefit from a longer-term treatment for teriparatide

Table 2 Definition of very high risk of fracture American association of clinical endocrinology guidelines 2020, endocrine society guidelines 2020

American association of clinical endocrinology guidelines 2020	Endocrine society guidelines 2020
Very high risk • Recent fracture in the past 12 months • Multiple fractures • Fractures while on approved osteoporosis therapy • Fractures while on drugs causing skeletal harm • High risk for falls or history of injurious falls • A very low T score (<−3.0) • A very high fracture probability by fracture risk assessment tool (major osteoporotic fracture >30%, hip fracture >4.5%)	Very high risk • Multiple spine fractures and a bone mineral density at the hip or spine of ≤−2.5 High risk • T-score of ≤−2.5 • Prior hip or vertebral fracture • High fracture risk as calculated by fracture assessment tool (10-year probability of major osteoporotic fracture ≥20%, hip fracture ≥3%

are patients with elevated P1NP after two years of treatment. This may be because elevated P1NP may signify that new bone formation is ongoing [16].

7.2 Abaloparatide

While the black box warning for osteosarcoma has been removed from the FDA label, more than 24 months of treatment is not recommended due to the lack of data evaluating safety beyond three years.

7.3 Transitioning to Antiresorptive Treatment After Teriparatide and Abaloparatide

The benefit and utility of transitioning to an anti-resorptive agent after the completion of anabolic therapy are essential. Data clearly shows that after teriparatide is discontinued, there is a rapid decline in bone density if an active antiresorptive medication is not used. While no data is available for abaloparatide, a similar BMD decrease will likely occur if not followed by anti-remodeling treatment [15]. Therefore, a potent anti-resorptive/remodeling medication such as bisphosphonate and denosumab is recommended after teriparatide or abaloparatide treatment. Notably, further gains in bone mineral density and continued fracture risk reduction are observed when patients' transition to antiresorptive/remodeling treatment after teriparatide or abaloparatide [15]. For example, in the ActiveExtend trial, the sustained fracture risk reduction was noted with further gains in bone mineral density with 18 months of abaloparatide therapy followed by 24 months of alendronate treatment [21].

7.4 Transitioning to Teriparatide After Antiresorptive Treatment

Patients who receive teriparatide after bisphosphonates such as alendronate or risendronate have blunted bone mineral density increases as compared to receiving teriparatide only [22]. In the DATA switch study, patients who received teriparatide 2 years followed by 2 years denosumab, had further gains in bone mineral density [23]. However, in patients who received 2 years of teriparatide after denosumab, BMD in the femoral neck decreases in the first 12 months of receiving denosumab. This was associated with increased in cortical porosity (as compared to teriparatide alone) and decrease in estimated strength of bone by finite element analysis [24]. It is increasingly being recognized that if possible osteoanabolic treatment are used before antiresorptive treatments. Transitioning from denosumab to teriparatide may be less than ideal, due to previously mentioned bone loss, with romosozumab being a better option in this setting.

8 Practical Considerations for the Use of PTH and PTHRP Analogs

Teriparatide (prefilled syringe with 28 doses) is given as a subcutaneous injection in the thigh or abdomen. Abaloparatide (prefilled syringe with 30 doses) is provided as a subcutaneous injection in the abdomen. For abaloparatide to have a better effect, it is recommended for administration in the periumbilical area of the abdomen because this area allows for more rapid absorption of the drug due to higher levels of circulation [9]. These injections are self-administered by the patient. While the injection device of teriparatide needs to be always refrigerated, the injection device of abaloparatide can be kept at room temperature after opening. Due to the possibility of dizziness and orthostatic hypotension, performing the subcutaneous injections while the patient is sitting is prudent. While the FDA recommendation does not mention the requirement for monitoring for hypercalcemia, many clinicians do feel it prudent to measure serum calcium with teriparatide use. A commonly used clinical protocol for monitoring is to check predose serum calcium one month and three months after starting teriparatide [16]. For abaloparatide, no laboratory monitoring is recommended except for patients at risk for hypercalcemia or renal stones [15].

9 Summary

The development of teriparatide (PTH analog) and abaloparatide (PTHrP analog) has expanded our armamentarium to treat osteoporosis. Teriparatide and abaloparatide are pro-remodeling osteoanabolic agents that can significantly gain bone mineral density while reducing fracture risk. Abaloparatide and teriparatide are most helpful in patients at high risk for fractures. They can be combined with antiresorptive agents to treat osteoporosis and provide maximum benefit to patients.

References

1. Lim SY, Bolster MB. Current approaches to osteoporosis treatment. Curr Opin Rheumatol 27:216–224
2. Silva BC, Costa AG, Cusano NE et al (2011) Catabolic and anabolic actions of parathyroid hormone on the skeleton. J Endocrinol Invest 34:801–810
3. Brent MB (2021) Abaloparatide: a review of preclinical and clinical studies. Eur J Pharmacol 909:174409. https://doi.org/10.1016/j.ejphar.2021.174409
4. Shirley M (2017) Abaloparatide: first global approval. Drugs 77:1363–1368
5. Lindsay R, Nieves J, Formica C et al (1997) Randomised controlled study of effect of parathyroid hormone on vertebral-bone mass and fracture incidence among postmenopausal women on oestrogen with osteoporosis. Lancet 350:550–555
6. Silva BC, Bilezikian JP (2015) Parathyroid hormone: anabolic and catabolic actions on the skeleton. Curr Opin Pharmacol 22:41–50
7. Dempster DW, Zhou H, Ruff VA et al (2018) Longitudinal effects of teriparatide or zoledronic acid on bone modeling- and remodeling-based formation in the SHOTZ study. J Bone Miner Res 33:627–633
8. United States Food and Drug Administration (2020) FORTEO (teriparatide) Label. https://www.accessdata.fda.gov/drugsatfda_docs/label/2021/021318Orig1s056lbl.pdf. Accessed 19 February 2023
9. United States Food and Drug Administration (2021) TYMLOS® (abaloparatide) injection, for subcutaneous use. https://www.accessdata.fda.gov/drugsatfda_docs/label/2021/208743s010lbl.pdf. Accessed 19 February 2023
10. Neer RM, Arnaud CD, Zanchetta JR et al (2001) Effect of parathyroid hormone (1–34) on fractures and bone mineral density in postmenopausal women with osteoporosis. N Engl J Med 344:1434–1441
11. Miller PD, Hattersley G, Riis BJ et al (2016) Effect of abaloparatide vs placebo on new vertebral fractures in postmenopausal women with osteoporosis: a randomized clinical trial. JAMA 316:722–733
12. Cosman F (2021) Teriparatide and abaloparatide treatment for osteoporosis. In: Dempster DW, Cauley JA, Bouxsein ML et al (eds) Marcus and Feldman's osteoporosis, 5th edn. Academic Press, Massachusetts, pp 1757–1769
13. Saag KG, Zanchetta JR, Devogelaer JP et al (2009) Effects of teriparatide versus alendronate for treating glucocorticoid-induced osteoporosis: thirty-six-month results of a randomized, double-blind, controlled trial. Arthritis Rheum 60:3346–3355
14. Saag KG, Shane E, Boonen S et al (2007) Teriparatide or alendronate in glucocorticoid-induced osteoporosis. N Engl J Med 357:2028–2039
15. McClung MR (2021) Role of bone-forming agents in the management of osteoporosis. Aging Clin Exp Res 33:775–791
16. Miller PD, Lewiecki EM, Krohn K et al (2021) Teriparatide: label changes and identifying patients for long-term use. Cleve Clin J Med 88:489–493
17. Hodsman AB, Bauer DC, Dempster DW et al (2005) Parathyroid hormone and teriparatide for the treatment of osteoporosis: a review of the evidence and suggested guidelines for its use. Endocr Rev 26:688–703
18. Vahle JL, Sato M, Long GG et al (2002) Skeletal changes in rats given daily subcutaneous injections of recombinant human parathyroid hormone (1–34) for 2 years and relevance to human safety. Toxicol Pathol 30:312–321

19. Gilsenan A, Midkiff K, Harris D et al (2021) Teriparatide did not increase adult osteosarcoma incidence in a 15-year us postmarketing surveillance study. J Bone Miner Res 36:244–251
20. Bilezikian JP, Hattersley G, Mitlak BH et al (2019) Abaloparatide in patients with mild or moderate renal impairment: results from the ACTIVE phase 3 trial. Curr Med Res Opin 35:2097–2102
21. Bone HG, Cosman F, Miller PD et al (2018) ACTIVExtend: 24 months of alendronate after 18 months of abaloparatide or placebo for postmenopausal osteoporosis. J Clin Endocrinol Metab 103:2949–2957
22. Cosman F, Nieves JW, Dempster DW (2017) Treatment sequence matters: anabolic and antiresorptive therapy for osteoporosis. J Bone Miner Res 32:198–202
23. Leder BZ, Tsai JN, Uihlein AV et al (2015) Denosumab and teriparatide transitions in post-menopausal osteoporosis (the DATA-Switch study): extension of a randomised controlled trial. Lancet 386:1147–1155
24. Tsai JN, Nishiyama KK, Lin D et al (2019) Effects of Denosumab and Teriparatide transitions on bone microarchitecture and estimated strength: the DATA-Switch HR-pQCT Study. J Bone Miner Res 34:976

Medication-Related Osteonecrosis of the Jaw

Leanne Teoh, Michael McCullough, and Mathew Lim

Abstract The use of antiresorptive therapies, such as bisphosphonates and denosumab, has helped to significantly advance the management of osteoporosis. The introduction, and eventual widespread use of these medications soon led to increasing reports of a new adverse event called medication-related (previously bisphosphonate-related) osteonecrosis of the jaw (MRONJ). MRONJ remains a relatively rare condition associated with the use of antiresorptive agents for osteoporosis and is characterized by exposed areas of necrotic jaw bone. These areas of necrotic bone may range from an asymptomatic ulcer in the mouth, to debilitating presentations complicated by pain and infection that may progress to involve oral-cutaneous fistulae and predispose to pathologic fractures of the mandible. Recognition of the potentially significant impact on the quality of life of this condition has resulted in increased efforts to ensure screening of oral health and necessity for dental treatments, particularly invasive procedures, such as dental extractions, prior to commencement of these therapies. This chapter will provide an overview of the current understanding of MRONJ in relation to incidence, pathophysiology, risk factors and clinical presentation. In addition, it will provide a brief overview of approaches to prevention and management of this condition.

Keywords Antiresorptive therapies · Bisphosphonate-related osteonecrosis of the jaw · Medication-related osteonecrosis of the jaw · Bisphosphonates · Denosumab · Romosozumab · Glucocorticosteroids · Anti-angiogenic therapies

1 Introduction

Medication-related osteonecrosis of the jaw (MRONJ) is an uncommon but potentially debilitating condition characterized by necrotic, exposed jaw bone, and is associated with pain, infection, soft tissue abscesses, sinusitis, draining fistulae and significantly affects the patient's quality of life [1, 2]. MRONJ can occur spontaneously,

L. Teoh (✉) · M. McCullough · M. Lim
Melbourne Dental School, The University of Melbourne, Carlton, VIC, Australia
e-mail: leanne.teoh@unimelb.edu.au

© The Author(s), under exclusive license to Springer Nature Singapore Pte Ltd. 2023   61
A. H. Choi and S. Yik Lim (eds.), *Pharmacological Interventions for Osteoporosis*,
Tissue Repair and Reconstruction, https://doi.org/10.1007/978-981-99-5826-9_5

but is most frequently associated with an invasive dental procedure, most commonly a tooth extraction [2].

The first report of MRONJ was by Marx in 2003 who described this adverse event associated with bisphosphonates [3]. Since then, a plethora of literature has been published associating it with other drugs that affect bone turnover and wound healing, including denosumab, romosozumab, and anti-angiogenic drugs. Due to the increased dose and frequency of bisphosphonates and denosumab for patients with cancer, and coupled with the patient's underlying medical condition, the incidence of MRONJ is higher for oncology patients. Nevertheless, these medicines carry a lower but significant risk when used for osteoporosis.

2 Definition

MRONJ is defined by the American Society for Bone and Mineral Research and listed in the recent position paper by the American Academy of Oral and Maxillofacial Surgeons (AAOMS) as having the following three characteristics [4, 5]:

(1) Current or previous treatment with antiresorptive therapy alone or in combination with immune modulators or antiangiogenic medications.
(2) Exposed bone or bone that can be probed through an intraoral or extraoral fistula(e) in the maxillofacial region that has persisted for more than 8 weeks.
(3) No history of radiation therapy to the jaws or metastatic disease to the jaws.

This definition allows the clinician to exclude different forms of osteonecrosis of the jaw, such as cases that occur spontaneously, or that are associated with head/neck radiation therapy.

3 Incidence and Risk

The incidence of MRONJ differs depending on whether the bisphosphonates or denosumab are used for an oncology patient, or a patient with osteoporosis. This is due to the much higher doses of anti-resorptive agents used in cancer patients to prevent or treat bony metastases or prevent skeletal complications. The International Task Force of Osteonecrosis of the Jaw reported the incidence of MRONJ is between 1 and 15% in patients taking these medicines for cancer, and between 0.001 and 0.01% for patients taking associated medicines for osteoporosis [6].

The risk of MRONJ is between 0.02 and 0.05% for patients taking bisphosphonates for osteoporosis [5], with the risk for patients exposed to intra-venous (IV) zoledronate being $\leq 0.02\%$, and those taking oral bisphosphonates being $\leq 0.05\%$, respectively [5]. The risk of developing MRONJ for a patient taking denosumab for osteoporosis is between 0.04 and 0.3% [5]. Lastly, the risk of developing MRONJ for a patient using romosozumab is estimated to be between 0.03 and 0.05% [5].

It is important to note that the duration of therapy also affects risk, with durations of use of bisphosphonates four years or more carrying a higher degree of risk of MRONJ [7].

4 Pathophysiology

The pathophysiology of MRONJ is unknown, but centered around five predominant theories, including altered bone turnover, localized infection or inflammation, altered angiogenesis, immune dysfunction and genetic predisposition [5, 8]. However, it likely that the development and progression of MRONJ is multifactorial, influenced by dental, medical and medication-related factors [8].

Osteoclast differentiation and function are critical for bone turnover and remodeling processes. As bisphosphonates and denosumab inhibit osteoclast function, inhibition of bone remodeling is one method by which it is proposed that these drugs are associated with MRONJ [8].

It is established that the presence of localized dental infection or inflammation predisposes to MRONJ. While the vast majority of MRONJ cases are precipitated by a tooth extraction, most of these extractions would have been conducted due to the presence of an odontogenic or periodontal infection. Evaluation of histological sections of MRONJ in animal models have detected the presence of inflammatory cytokines [8, 9]. In patients taking bisphosphonates for bony metastases due to solid tumors, the intervention of dental hygiene (deep oral hygiene treatments and professional root planning and scaling, and extraction of teeth with a poor prognosis) prior to commencement of bisphosphonates led to a reduction in MRONJ [10]. It is still unclear whether the bacteria precede the development of MRONJ, or caused by the biofilm developed after the necrotic bone is established [8].

Angiogenesis occurs as part of the bone remodeling and healing process after procedures such as a tooth extraction. In areas of MRONJ, animal models have shown a decreased arteriole and venule vasculature, together with a reduction in other markers of angiogenesis including vascular endothelial growth factor-A [11]. Soft tissue toxicity has also been proposed to be a contributing factor to the development of MRONJ, with altered cellular and tissue responses including oxidative stress, hypoxia and apoptosis shown in animal models [11]. More recently, several immunomodulatory and antiangiogenic drugs, including tumor necrosis factor inhibitors, cytotoxic chemotherapy agents and mammalian target of rapamycin inhibitors have been proposed to be associated with the development of MRONJ, highlighting the complex multifactorial pathophysiological process [12, 13].

Further, innate or acquired immune dysfunction has been implicated in the pathophysiology of MRONJ. Patients with reduced immune function due to diabetes, kidney transplant and some autoimmune conditions are at increased risk of developing MRONJ [2, 14]. In patients who are taking antiresorptive medicines, the use of glucocorticosteroids and other immunosuppressants are shown to be additional risk factors for MRONJ. However, immunosuppression in isolation, with no other risk factors, is not associated with MRONJ, as hematological markers of a reduced immune system are not predictive of MRONJ risk [15].

Lastly, there is a weak evidence to show that genetic predilection may exist, with single-nucleotide polymorphisms (SNPs) associated with the genes that encode for processes affecting bone turnover and angiogenesis, including vascular endothelial growth factor. These SNPs have been shown to be present in patients with both multiple myeloma and those treated with bisphosphonates for osteoporosis [5, 16, 17]. Further research is required to determine if there is an increased risk for this cohort of patients [5].

5   Risk Factors

MRONJ is multifactorial in nature, with dental, medication or systemic risk factors that need to be considered.

5.1   Dental Risk Factors

MRONJ is most commonly associated with a dental extraction, although there are cases of MRONJ developing with no obvious precipitating factor [2]. A retrospective study to determine the frequency and clinical characteristics of MRONJ in Australia showed that a tooth extraction was the precipitating factor in 73% of cases [18]. A systematic review of risk factors of MRONJ also determined that tooth extraction was the most common trigger [14]. However, other dental procedures are also implicated, including periodontal disease management, denture trauma and implant placement [7, 14].

Anatomical factors are also a consideration, with MRONJ more likely to develop in the mandible compared to the maxilla potentially due to the differences in vascularity [7] (Fig. 1).

5.2   Medical Risk Factors

Various medical conditions are associated with an increased risk of MRONJ, partly due to the condition itself as well as the medications used to treat these conditions. Patients with cancer, particularly multiple myeloma, breast cancer, prostate cancer and renal cancer have clearly been demonstrated to have an increased risk, although other cancers including lung, leukemia and thyroid have also been associated with MRONJ [14] (Fig. 2). This is due to the immunocompromised condition of the patient, coupled with the high doses and frequency of the antiresorptive medicines used to treat bony metastases or prevent skeletal fractures. Patients with osteoporosis

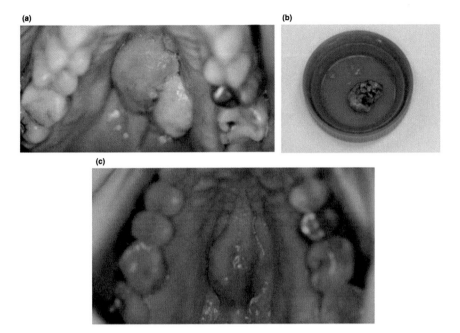

Fig. 1 A 66-year-old female with a history of osteoporosis, managed with oral alendronate, presented with stage 1 MRONJ of her palatal torus following burning the roof of her mouth while eating (**a**). The patient was managed with non-operative measures but eventually developed symptoms of discomfort with a larger area of exposed bone. The area of necrotic bone was loose and a sequestrectomy completed. **b** shows the removed bone fragment from the palatal torus. **c** The palatal torus 4-weeks following the sequestrectomy

are at risk due to the antiresorptive medicines used. Other medical conditions are associated with MRONJ, such as kidney transplant, rheumatoid arthritis and Paget's disease [14]. This is likely to be due both to the antiresorptive and antiangiogenic medicines used to manage these conditions as well as the pathophysiological disease processes involved. The International Task Force on Osteonecrosis of the Jaw further identified tobacco use, diabetes, hyperthyroidism, renal dialysis and increasing age as significant systemic factors that increase an individual's risk of developing MRONJ [6].

5.3   Medication Risk Factors

5.3.1   Bisphosphonates

Bisphosphonates were the first medication associated with MRONJ. The proposed mechanism is via suppression of osteoclast-mediated bone resorption through

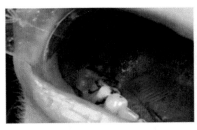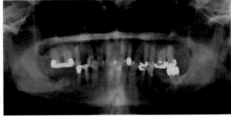

Fig. 2 A 69-year-old male with metastatic prostate cancer with stage 1 MRONJ. The patient was being managed on various trials due to his recalcitrant disease which included monthly doses of denosumab. (Left) is taken following exfoliation of the mandibular first molar, tooth 46. The adjacent molar, tooth 47, had exfoliated a year earlier. (Right) The panoramic radiograph was taken the day the tooth 46 exfoliated and shows the possible extent of the MRONJ in the right mandible

inducing osteoclast apoptosis after the bisphosphonate binds to and inhibits the action of the regulatory enzyme farnesyl pyrophosphate synthase. It has been suggested that bisphosphonates also inhibit angiogenesis by inhibition of VEGF expression and reduction of other mediators involved in angiogenesis, including FGF-2 and MMP-2 [19]. Thus, after procedures such as a tooth extraction, the ability of the bone to turnover and heal is reduced for patients taking bisphosphonates. The indication for therapy also affects MRONJ risk, with these medicines used in oncology patients, places them at significantly higher risk compared to those using it for osteoporosis, due to the much higher doses and frequency of administration of bisphosphonate. The extremely high affinity of bisphosphonates for bone matrix, with an estimated bone half-life of alendronate being 10 years, has resulted in an accepted risk of bisphosphonates causing MRONJ to be life-long [20]. It is for this reason that drug holidays of bisphosphonates have not been recommended to prevent MRONJ for patients who need dental procedures while taking these medicines [21].

5.3.2 Denosumab

The dose and indication for use of denosumab also affects MRONJ risk, with cancer patients who receive much higher and more frequent doses of denosumab at increased risk. Denosumab is thought to be associated with MRONJ due to inhibition of RANKL. By inhibiting the activation of osteoclasts, denosumab subsequently impairs osteoclast-mediated bone resorption and bone turnover. The interception of the RANK-RANKL binding could also affect monocyte and macrocyte function, which is another proposed mechanism for denosumab associated MRONJ risk [19]. While patients taking denosumab for osteoporosis may be at a slightly greater risk of MRONJ compared to those taking oral bisphosphonates, patients with a history of bisphosphonate therapy prior to taking denosumab for osteoporosis are at increased risk for MRONJ [22], likely due to the prolonged bone half-life of bisphosphonates.

5.3.3 Romosozumab

While romosozumab is a relatively new anabolic agent for the treatment of osteoporosis with the dual action of inhibiting bone resorption and inducing bone formation, it has been associated with two cases of MRONJ [23, 24]. While the incidence is estimated to be similar to bisphosphonates, further research and monitoring for adverse events are required to determine the strength of association between romosozumab and MRONJ.

5.3.4 Glucocorticosteroids

It has been established that the use of glucocorticosteroids is associated with increased risk of MRONJ when prescribed concurrently with anti-resorptive agents [5, 6]. This is likely due to their well-known effects on bone turnover that includes reduction in osteoblast numbers and proliferation, protein synthesis by osteoblasts and thus bone formation, as well as increased apoptosis of osteoclasts [25]. Additionally, glucocorticoids affect wound healing with a decreased infiltration and activation of inflammatory cytokines, growth factors and matrix proteases, as well as inhibiting collagen matrix production [26].

5.3.5 Other Medications Associated with MRONJ

Several other medicines, including tumor necrosis factor inhibitors and anti-angiogenics have been implicated in MRONJ, particularly when used together with antiresorptives. Nevertheless, there are case reports of these medicines associated with MRONJ when used in isolation [13]. The TNF-alpha inhibitor adalimumab has had several case reports associating it with MRONJ [27, 28]. This human monoclonal antibody is thought to inhibit bone turnover by either diminishing RANKL, or increasing apoptosis of monocytes, and therefore diminish the ability of the bone to heal after procedures such as tooth extractions [27].

Drugs that affect angiogenesis, including bevacizumab, VEGF receptor blockers such as tyrosine kinase inhibitors sunitinib and sorafenib, and mammalian target of rapamycin inhibitors everolimus and sirolumus, also have several cases reports in association with MRONJ [13, 29, 30]. It is thought that these drugs intercept wound healing by impairing angiogenesis in the jawbone after procedures, such as tooth extractions.

Many other drugs have been implicated as contributors to the development of MRONJ in case studies, such as mycophenolate, methotrexate, cyclophosphamide, docetaxel and thalidomide [12]. While case studies are the lowest level of evidence, it is important to be aware that MRONJ is a relatively newly established condition and the list of drugs associated with it is increasing. Given that MRONJ is a rare adverse effect, this underscores the importance of reporting suspected cases of MRONJ to pharmacovigilance bodies.

6 Prevention

It is recommended to refer the patient to a dentist prior to the initiation of antiresorptive therapy, to ensure the patient has optimal oral health prior to the commencement of therapy. Dentists should undertake a comprehensive oral examination both clinically and using radiographs, and have the patient ideally dentally fit (treating any caries, root fragments, broken teeth, periodontal or periapical pathology) and ensuring dentures are well fitting [5, 31]. Any infection should be treated appropriately, and any required extractions should be undertaken. Several studies have shown the value in performing any required high risk surgical procedures prior to initiating antiresorptive therapy to minimize the risk of MRONJ [10, 31, 32].

Informing the patient of the risks of MRONJ and importance of regular dental attendance for ongoing, maintenance care is imperative for the patient to maintain their oral health to minimize the need for invasive procedures as well as to treat any local infection or inflammation. Encouraging smoking cessation and educating patients about signs and symptoms of MRONJ is also important for MRONJ prevention.

If the patient is already on antiresorptive therapy and requires an invasive dental procedure such as an extraction, various local measures have been recommended including soft tissue closure, where possible, by suturing the socket and reviewing the patient 8 weeks after the procedure to check for exposed bone [21]. Pre-procedural mouthwashes such as chlorhexidine, as well as prophylactic antibiotics have been recommended in some guidelines with varying degrees of evidence [21, 32, 33]. Table 1 shows MRONJ prevention strategies.

6.1 Drug Holidays

In the past, drug holidays of antiresorptive medicine have been recommended to minimize the risk of MRONJ. However, due to the high affinity for bone for bisphosphonates and their prolonged bone half-life, it is not recommended to cease bisphosphonates as this will have little benefit for MRONJ prevention. Additionally, the evidence that demonstrates the benefit of a drug holiday from bisphosphonates for MRONJ prevention is inconclusive [5].

The cessation of denosumab for osteoporosis to minimize MRONJ risk is not recommended. Unlike bisphosphonates, denosumab does not bind to bone and their effects on bone turnover are minimal 6 months after the last dose [5]. Population-based cohort studies and several reviews of denosumab have demonstrated that bone density rapidly declines to baseline and the risk of vertebral fractures is increased 4–16 weeks after cessation of denosumab, and as such, it is not recommended to cease denosumab [34–36]. Dental procedures are recommended to be performed around 8 weeks prior to the next dose, while denosumab plasma concentrations are declining to minimize any risk of MRONJ [5].

Table 1 MRONJ prevention strategies. Reprint with permission from [5]

Pretherapy (nonmalignant disease)	• Educate patient about the potential risks associated with long-term ART[a] • Optimization of dental health can occur concurrent with ART
Pretherapy (malignant disease)	• Educate patients about the higher risks of MRONJ and the importance of regimented dental care • Optimization of the dental health prior to the initiation of ART if systemic conditions permit (extraction of non-restorable teeth or teeth with a poor prognosis)
During antiresorptive therapy (nonmalignant disease)	• No alteration of operative plan for most patients • Considerations include drug schedule, duration of therapy, comorbidities, other medications (especially chemotherapy, steroids, or antiangiogenics), degree of underlying infection/inflammation, and extent of surgery to be performed. Drug holidays are controversial • BTM[b] are not a useful tool to assess MRONJ risk
During antiresorptive therapy/targeted therapies (malignant disease)	• Educate patients about the higher MRONJ risk in the setting of malignant disease • Educate the patient about the importance of regimented dental care and prevention • Avoid dentoalveolar surgery if possible • Consider root retention techniques to avoid extractions • Dental implants are contraindicated • Drug holidays are controversial

[a] Antiresorptive therapies
[b] Bone turnover markers (CTX)

6.2 Biomarkers to Predict MRONJ Risk

Various biomarkers of bone turnover have been trialed to determine if they are predictive of a patient's risk of MRONJ. In the past, C-terminal telopeptide (CTX), the product of type 1 collagen produced during bone resorption, had been used to determine if specific serum concentrations would be indicative of risk. However, this has been shown not to be an accurate predictor, given that CTX levels are influenced by many factors, including age, alcohol consumption, circadian rhythms and other medications [37, 38]. There are currently no biomarkers that are recommended to aid in determining MRONJ risk.

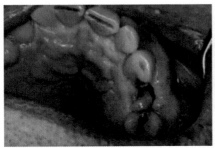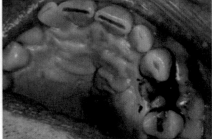

Fig. 3 A case of a 57-year-old male with history of multiple myeloma. Part of his medical treatment included an extensive history of zoledronic acid use. He underwent an extraction of a symptomatic and unrestorable maxillary left first premolar (tooth 24). (Left) The non-healing and asymptomatic extraction socket at 8 weeks post-extraction represents a diagnosis of stage 1 MRONJ. (Right) The extraction socket immediately post-sequestrectomy completed at 12 weeks. Removal of the necrotic bone fragments revealed the socket to be epithelialized. Healing progressed uneventfully after removal of the necrotic bone

7   Clinical Presentation

7.1   Clinical Findings

The clinical appearance of MRONJ is highly variable. Given the association with invasive oral surgical procedures, the most expected and recognizable presentation is a non-healing extraction socket, with limited to no gingival coverage and visible exposed yellow-white necrotic bone, with either smooth or irregular borders, similar to that of alveolar osteitis (dry socket) (Fig. 3). Many patients at risk of MRONJ will have comorbidities that predispose to delayed healing, thus justifying the case definition that requires exposed bone to be present for a minimum of 8 weeks [39]. The presence of local inflammation, infection, or pain will vary between patients and assist with defining staging and management [5].

MRONJ may also present as a bone sequestrum, draining sinus or fistula to bone, or an ulcer with a base of exposed necrotic bone. Areas of the mouth predisposed to mucosal trauma, such as prominent bony anatomical features (e.g. exostoses, particularly lingual or palatal tori, mylohyoid ridge) and areas supporting removable denture prostheses are most commonly affected by the so-called 'spontaneous' cases [39].

7.2   Symptoms

An estimated two-thirds of MRONJ cases are reported to have associated pain, however, the nature and severity of symptoms is highly variable [40]. In many cases, MRONJ lesions will be completely asymptomatic, with pain caused primarily by secondary trauma from sharp bone to other soft tissues (Fig. 4) [7]. Pain from MRONJ lesions is often attributed to infection of the necrotic tissues and inflammation of the adjacent vital tissues. In most cases, pain is of a dull or aching quality with mild to moderate intensity, but this may increase to severe when irritated [40]. Despite a general lack of association between lesion size and pain intensity, more extensive lesions have a tendency to develop chronic, neuropathic pain.

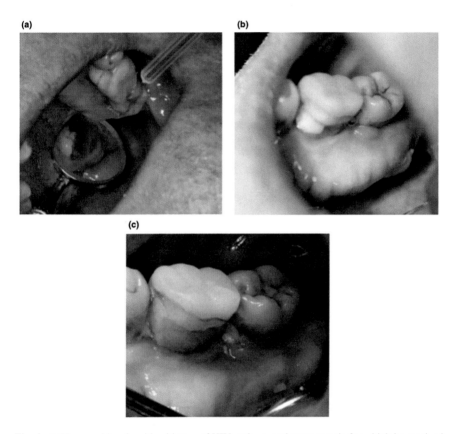

Fig. 4 A 55-year-old male with a history of HIV, asthma, and osteoporosis for which he received two doses of zoledronic acid. The patient presented with spontaneous MRONJ of the mylohyoid ridge, which was only noticed following traumatic ulceration of the tongue. **a** The area of exposed bone with associated gingival inflammation, representing stage 2 MRONJ. **b** The area of exposed bone after two weeks, following smoothing of the exposed bone and oral hygiene with chlorhexidine mouth rinses, was no longer symptomatic, moving to stage 1 MRONJ. **c** Over several months, the necrotic bone exfoliated and eventually healed

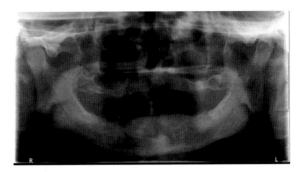

Fig. 5 A 70-year-old male taking denosumab for the management of osteoporosis secondary to a complex list of comorbidities and chronic gastroparesis. The patient presented with symptoms of soreness in the right mandible that did not appear to be related to his full lower denture. The panoramic radiograph demonstrated changes associated with stage 0 MRONJ which eventually progressed to exposure of necrotic bone in the area

7.3 Radiographic Changes

Radiographic changes associated with MRONJ may be subtle and non-specific [41]. Commonly reported changes include a lack of bony infill in extraction sockets, alveolar bone loss or resorption not able to be attributed to other causes such as periodontal disease (Fig. 5). In addition, there can be changes to the bone trabecular pattern (moth-eaten appearance) and periodontal ligament space (thickened lamina dura and decreased periodontal ligament space, sclerotic lesions) [5]. Due to the subtle nature of these changes, they may only be noticed on close comparison of standardized radiographs to those completed or pre-treatment assessments (Fig. 6). As a result, although primary radiographic examinations may be of use, there is limited benefit in additional radiographic investigations, such as magnetic resonance imaging or nuclear medicine, except in cases of doubtful diagnosis [41]. 3D imaging modalities may be of benefit in more severe cases where MRONJ lesions are suspected to extend beyond the alveolar ridge or place the integrity of the involved jawbone at risk.

8 Staging

Various staging systems for MRONJ have been proposed, classifying the condition severity based on clinical and/or radiographic features, symptoms, and indications for medical or surgical intervention [5, 42–44].

While no staging system is recognized universally, the most commonly used is that proposed by the AAOMS initially in 2007, and then later modified in 2009 (Table 2) [39, 45]. The AAOMS proposed classification system is based on the clinical detection of exposed necrotic bone, or a fistula able to probe necrotic bone (Fig. 7) [5, 7]. This differentiates between stage 1 lesions and precursor conditions,

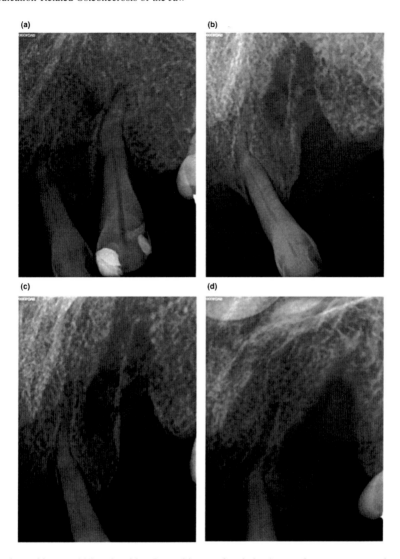

Fig. 6 An 83-year-old female with a 5-year history of oral alendronate for management of osteoporosis. She had a symptomatic and mobile left maxillary canine (tooth 23) extracted in preparation for a new upper denture. The series of radiographs shows common radiographic changes associated with MRONJ: The pre-operative radiograph of the tooth 23 (**a**). Periapical radiographs taken at 4 weeks (**b**) and 8 weeks (**c**) of the extraction socket when a fistula was noted at the crest of the alveolus at the extraction site. Despite these radiographs demonstrating a lack of bony infill into the extraction socket, this would not necessarily be expected in this timeframe. There is a distinct 'moth-eaten appearance' of the socket itself and to the mesial of the socket demonstrating the extent of the MRONJ. **d** is taken 4 weeks post-debridement of the MRONJ following healing of the extraction socket clinically

such as stage 0 MRONJ, that were added to reflect suspected prodromal changes, symptomatically or radiographically, prior to exposure of the necrotic bone. More advanced lesions are defined by symptoms and the presence of inflammation and/or infection (Stage 2) and the extent of lesions that may compromise the integrity of the involved bone (Stage 3) (Fig. 8). The most recent AAOMS position paper now proposes possible treatment interventions based on this staging [5].

Table 2 AAOMS staging of medication-related osteonecrosis of the jaw [7]

Stage	Description	Symptoms	Clinical findings	Radiographic findings
Patient at risk	No apparent necrotic bone in an asymptomatic patient exposed with antiresorptive therapy	Asymptomatic	No signs of necrotic bone	No radiographic changes
Stage 0 (non-exposed bone variant)	No clinical evidence of necrotic bone, but non-specific symptoms or clinical and radiographic findings. May occur in patients with prior history of stage 1–3 disease who have healed, or may be potential precursor to stage 1 disease	Odontalgia not explained by another cause. Dull, aching jaw bone pain that may radiate to temporomandibular joint region. Sinus pain, which may be associated with inflammation and thickening of the maxillary sinus wall. Altered neurosensory function	No visible necrotic bone or fistulae. No inflammation or infection. Loosening of teeth not explained by periodontal disease. Intraoral or extraoral swelling	Alveolar bone loss or resorption not attributable to periodontal disease. Changes to trabecular pattern sclerotic bone and no new bone in extraction sockets. Regions of osteosclerosis involving the alveolar bone and/or surrounding basilar bone. Thickening/obscuring of the periodontal ligament (thickened lamina dura, sclerosis, decreased size of periodontal ligament space)

(continued)

Table 2 (continued)

Stage	Description	Symptoms	Clinical findings	Radiographic findings
Stage 1	Exposed necrotic bone, or fistula that probes to bone, in asymptomatic patients with no evidence of inflammation or infection	Asymptomatic	Exposed necrotic bone or area or necrotic bone that can be probed through fistula No inflammation or infection	May present with stage 0 radiographic changes localized to alveolar bone region
Stage 2	Symptomatic, exposed and necrotic bone, or fistula that probes to bone, with evidence of inflammation and/or infection	Symptomatic	Exposed necrotic bone or area or necrotic bone that can be probed through fistula Evidence of inflammation or infection	May present with stage 0 radiographic changes localized to alveolar bone region
Stage 3	Symptomatic, exposed and necrotic bone, or fistula that probes to bone, with evidence of infection and advanced clinical findings indicative of extensive disease	Symptomatic	Exposed necrotic bone or fistula with signs of infection and one of more of the following: – Exposed necrotic bone extending beyond the region of the alveolar bone (i.e. inferior border and ramus of mandible, maxillary sinus, zygoma) – Pathologic fracture – Extraoral fistula – Oral antral/ oronasal communication – Osteolysis extending to inferior border of mandible or sinus floor	Radiographic changes supportive of extensive disease changes e.g. pathologic fracture, osteolysis extending to inferior border of mandible or sinus floor

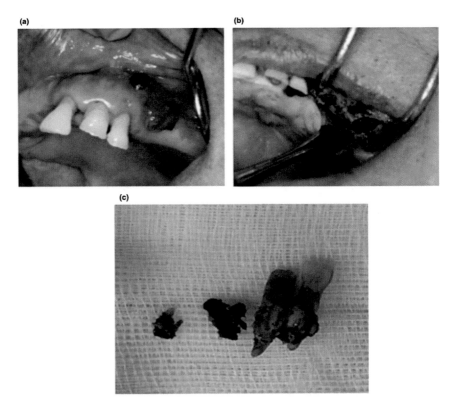

Fig. 7 A 75-year-old male with a complex medical history including polymyelitis, type 2 diabetes mellitus, cirrhosis secondary to hepatitis B, hypertension, hypercholesterolemia, long-term corticosteroid treatment, and osteoporosis managed with oral alendronate. The patient had a prior history of stage 2 MRONJ 5 years earlier requiring a partial maxillectomy and free-flap reconstruction. This is the patient's second presentation of stage 2 MRONJ with multiple draining sinuses in the previous unaffected left maxilla (**a**). Raising of a mucoperiosteal flap demonstrates the extent of the necrosis and the bleeding bone following debridement (**b**). The patient needed several debridement procedures and removal of the upper incisors, which were also involved in the extent of the MRONJ lesion and show attached necrotic bone (**c**)

9 Management

While prevention of MRONJ through the optimization and preservation of oral health is the ideal way, lesions may occur either spontaneously or as a result of necessary dental interventions. Once established, the aim of any therapy should be curative and to improve the quality of life. Management options currently fall into two main groups: non-operative and operative.

Non-operative options focus largely on patient education and reassurance, analgesia for control of pain, and management of any secondary infection with the aim to either stabilize or allow sequestration of necrotic bone. Despite the unpredictable

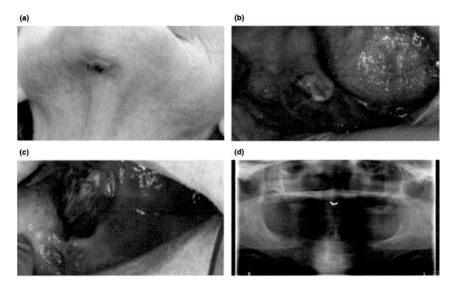

Fig. 8 An 89-year-old female with metastatic breast cancer, who received 16 doses of denosumab over the period of 2 years, presented with a non-healing submental lesion (**a**) and discomfort from her lower denture. The submental lesion was an extraoral fistula from a stage 3 MRONJ lesion in the right mandible (**b**). A second stage 2 MRONJ lesion was identified in the right maxilla (**c**). The panoramic radiograph (**d**) shows the extent of both lesions, especially the widespread changes in the anterior mandible

nature of progression of MRONJ lesions, resolution is possible in many early-stage cases. The measures recommended usually comprise of chlorhexidine mouth rinse and gels placed directly over the exposed bone to improve oral hygiene and prevent infection, with the addition of antibiotics if symptoms arise.

In cases where MRONJ is refractory or not responsive to non-operative measures, debridement of necrotic bone to vital, bleeding margins may be considered. While this has largely been used for more extensive lesions requiring segmental or marginal mandibulectomy or partial maxillectomy in the past, it may be used for all stages of MRONJ lesions, particularly where operative measures are seen to have potential to limit progression, induce healing, and improving quality of life. 3D imaging is of benefit to estimate the extent of the lesion and surgery.

Despite the relatively good response to operative measures, there remains the risk of refractory disease in compromised patients that have multiple risk factors. Guidelines remain uncertain about the evidence for a drug holiday to facilitate surgical interventions for MRONJ [5]. However, in recent clinical trials teriparatide promises the potential to offer a new adjunctive therapy in the management of refractory cases [46]. The evidence for other previously discussed therapies, including hyperbaric oxygen therapy, or vitamin E and pentoxifylline, are largely limited to case reports [5].

Ultimately, the foundation of MRONJ management should always be non-operative measures and close monitoring of progression given the often unpredictable course of lesions. The decision to use surgical measures or other adjunctive therapies should be patient and lesion specific, taking into consideration symptoms, the impact on quality of life, the ability to close the wound with soft tissues, morbidity of any intervention, and the potential for oral function and dental rehabilitation thereafter [5].

References

1. Kanwar N, Bakr MM, Meer M et al (2020) Emerging therapies with potential risks of medicine-related osteonecrosis of the jaw: a review of the literature. Br Dent J 228:886–892
2. Filleul O, Crompot E, Saussez S (2010) Bisphosphonate-induced osteonecrosis of the jaw: a review of 2,400 patient cases. J Cancer Res Clin Oncol 136:1117–1124
3. Marx RE (2003) Pamidronate (Aredia) and zoledronate (Zometa) induced avascular necrosis of the jaws: a growing epidemic. J Oral Maxillofac Surg 61:1115–1117
4. Khosla S, Burr D, Cauley J et al (2007) Bisphosphonate-associated osteonecrosis of the jaw: report of a task force of the American Society for Bone and Mineral Research. J Bone Miner Res 22:1479–1491
5. Ruggiero SL, Dodson TB, Aghaloo T et al (2022) American Association of Oral and Maxillofacial Surgeons' position paper on medication-related osteonecrosis of the jaws-2022 update. J Oral Maxillofac Surg 80:920–943
6. Khan AA, Morrison A, Kendler DL et al (2017) Case-based review of osteonecrosis of the jaw (ONJ) and application of the international recommendations for management from the international task force on ONJ. J Clin Densitom 20:8–24
7. Ruggiero SL, Dodson TB, Fantasia J et al (2014) American Association of Oral and Maxillofacial Surgeons position paper on medication-related osteonecrosis of the jaw–2014 update. J Oral Maxillofac Surg 72:1938–1956
8. Aghaloo T, Hazboun R, Tetradis S (2015) Pathophysiology of osteonecrosis of the jaws. Oral Maxillofac Surg Clin North Am 27:489–496
9. Aguirre JI, Akhter MP, Kimmel DB et al (2012) Oncologic doses of zoledronic acid induce osteonecrosis of the jaw-like lesions in rice rats (Oryzomys palustris) with periodontitis. J Bone Miner Res 27:2130–2143
10. Ripamonti CI, Maniezzo M, Campa T et al (2009) Decreased occurrence of osteonecrosis of the jaw after implementation of dental preventive measures in solid tumour patients with bone metastases treated with bisphosphonates. The experience of the National Cancer Institute of Milan. Ann Oncol 20:137–145
11. Gkouveris I, Hadaya D, Soundia A et al (2019) Vasculature submucosal changes at early stages of osteonecrosis of the jaw (ONJ). Bone 123:234–245
12. Teoh L, Moses G, Nguyen AP et al (2021) Medication-related osteonecrosis of the jaw: analysing the range of implicated drugs from the Australian database of adverse event notifications. Br J Clin Pharmacol 87:2767–2776
13. Pimolbutr K, Porter S, Fedele S (2018) Osteonecrosis of the jaw associated with antiangiogenics in antiresorptive-naïve patient: a comprehensive review of the literature. Biomed Res Int 2018:8071579. https://doi.org/10.1155/2018/8071579
14. McGowan K, McGowan T, Ivanovski S (2018) Risk factors for medication-related osteonecrosis of the jaws: a systematic review. Oral Dis 24:527–536
15. McGowan K, Ware RS, Acton C et al (2019) Full blood counts are not predictive of the risk of medication-related osteonecrosis of the jaws: a case-control study. Oral Surg Oral Med Oral Pathol Oral Radiol 128:373–380
16. Sarasquete ME, García-Sanz R, Marín L et al (2008) Bisphosphonate-related osteonecrosis of the jaw is associated with polymorphisms of the cytochrome P450 CYP2C8 in multiple myeloma: a genome-wide single nucleotide polymorphism analysis. Blood 112:2709–2712

17. Guo Z, Cui W, Que L et al (2020) Pharmacogenetics of medication-related osteonecrosis of the jaw: a systematic review and meta-analysis. Int J Oral Maxillofac Surg 49:298–309
18. Mavrokokki T, Cheng A, Stein B et al (2007) Nature and frequency of bisphosphonate-associated osteonecrosis of the jaws in Australia. J Oral Maxillofac Surg 65:415–423
19. Qi WX, Tang LN, He AN et al (2014) Risk of osteonecrosis of the jaw in cancer patients receiving denosumab: a meta-analysis of seven randomized controlled trials. Int J Clin Oncol 19:403–410
20. Drake MT, Clarke BL, Khosla S (2008) Bisphosphonates: mechanism of action and role in clinical practice. Mayo Clin Proc 83:1032–1045
21. Therapeutic Guidelines Limited (2019) Therapeutic guidelines-oral and dental, version 3. Therapeutic Guidelines Limited, Melbourne
22. Everts-Graber J, Lehmann D, Burkard JP et al (2022) Risk of osteonecrosis of the jaw under denosumab compared to bisphosphonates in patients with osteoporosis. J Bone Miner Res 37:340–348
23. Cosman F, Crittenden DB, Adachi JD et al (2016) Romosozumab treatment in postmenopausal women with osteoporosis. N Engl J Med 375:1532–1543
24. Saag KG, Petersen J, Brandi ML et al (2017) Romosozumab or alendronate for fracture prevention in women with osteoporosis. N Engl J Med 377:1417–1427
25. Reid IR (1998) Glucocorticoid effects on bone. J Clin Endocrinol Metab 83:1860–1862
26. Schäcke H, Döcke WD, Asadullah K (2002) Mechanisms involved in the side effects of glucocorticoids. Pharmacol Ther 96:23–43
27. Cassoni A, Romeo U, Terenzi V et al (2016) Adalimumab: another medication related to osteonecrosis of the jaws? Case Rep Dent 2016:2856926. https://doi.org/10.1155/2016/285 6926
28. Preidl RH, Ebker T, Raithel M et al (2014) Osteonecrosis of the jaw in a Crohn's disease patient following a course of Bisphosphonate and Adalimumab therapy: a case report. BMC Gastroenterol 14:6. https://doi.org/10.1186/1471-230X-14-6
29. Kim DW, Jung YS, Park HS et al (2013) Osteonecrosis of the jaw related to everolimus: a case report. Br J Oral Maxillofac Surg 51:e302–e304
30. Marino R, Orlandi F, Arecco F et al (2015) Osteonecrosis of the jaw in a patient receiving cabozantinib. Aust Dent J 60:528–531
31. Nicolatou-Galitis O, Schiødt M, Mendes RA et al (2019) Medication-related osteonecrosis of the jaw: definition and best practice for prevention, diagnosis, and treatment. Oral Surg Oral Med Oral Pathol Oral Radiol 127:117–135
32. Montefusco V, Gay F, Spina F et al (2008) Antibiotic prophylaxis before dental procedures may reduce the incidence of osteonecrosis of the jaw in patients with multiple myeloma treated with bisphosphonates. Leuk Lymphoma 49:2156–2162
33. Lodi G, Sardella A, Salis A et al (2010) Tooth extraction in patients taking intravenous bisphosphonates: a preventive protocol and case series. J Oral Maxillofac Surg 68:107–110
34. Lamy O, Stoll D, Aubry-Rozier B et al (2019) Stopping denosumab. Curr Osteoporos Rep 17:8–15
35. Lyu H, Yoshida K, Zhao SS et al (2020) Delayed denosumab injections and fracture risk among patients with osteoporosis: a population-based cohort study. Ann Intern Med 173:516–526
36. Pang KL, Low NY, Chin KY (2020) A Review on the role of denosumab in fracture prevention. Drug Des Devel Ther 14:4029–4051
37. Enciso R, Keaton J, Saleh N et al (2016) Assessing the utility of serum C-telopeptide cross-link of type 1 collagen as a predictor of bisphosphonate-related osteonecrosis of the jaw: a systematic review and meta-analysis. J Am Dent Assoc 147:551-560.e11
38. Awad ME, Sun C, Jernigan J et al (2019) Serum C-terminal cross-linking telopeptide level as a predictive biomarker of osteonecrosis after dentoalveolar surgery in patients receiving bisphosphonate therapy: systematic review and meta-analysis. J Am Dent Assoc 150:664-675.e8

39. Advisory Task Force on Bisphosphonate-Related Ostenonecrosis of the Jaws, American Association of Oral and Maxillofacial Surgeons (2007) American Association of Oral and Maxillofacial Surgeons position paper on bisphosphonate-related osteonecrosis of the jaws. J Oral Maxillofac Surg 65:369–376

40. Haviv Y, Geller Z, Mazor S et al (2021) Pain characteristics in medication-related osteonecrosis of the jaws. Support Care Cancer 29:1073–1080

41. Campisi G, Mauceri R, Bertoldo F et al (2020) Medication-related osteonecrosis of jaws (MRONJ) prevention and diagnosis: Italian consensus update 2020. Int J Environ Res Public Health 17:5998. https://doi.org/10.3390/ijerph17165998

42. Bedogni A, Fusco V, Agrillo A et al (2012) Learning from experience. Proposal of a refined definition and staging system for bisphosphonate-related osteonecrosis of the jaw (BRONJ). Oral Dis 18:621–623

43. Marx RE (2007) Oral and intravenous bisphosphonate-induced osteonecrosis of the jaws. Quintessence, Chicago

44. U.S. Department of Health and Human Services (2017) Common terminology criteria for adverse events (CTCAE) Version 5.0 2017. https://ctep.cancer.gov/protocoldevelopment/ele ctronic_applications/docs/CTCAE_v5_Quick_Reference_8.5x11.pdf. Accessed 06 Apr 2023

45. Ruggiero SL, Dodson TB, Assael LA et al (2009) American Association of Oral and Maxillofacial Surgeons position paper on bisphosphonate-related osteonecrosis of the jaws–2009 update. J Oral Maxillofac Surg 67:2–12

46. Sim IW, Borromeo GL, Tsao C et al (2020) Teriparatide promotes bone healing in medication-related osteonecrosis of the jaw: a placebo-controlled, randomized trial. J Clin Oncol 38:2971–2980

Fracture Liaison Service Optimization of Pharmacological Treatment for Osteoporosis Treatment

Joon-Kiong Lee and Hui-Min Khor

Abstract Optimization of pharmacological treatment for osteoporosis is important to reduce bone loss and risk of future fracture. Yet, many individuals who have suffered an osteoporosis fracture do not receive appropriate treatment for their bone health. Fracture Liaison Service (FLS) is designed to bridge the osteoporosis treatment gap. It is a coordinated programme involving a multidisciplinary team which identifies patients who sustained an index fracture, organizes bone health investigations, initiates intervention and provides follow-up monitoring of treatment. The implementation of FLS has led to increased rates of bone densitometry testing, initiation of osteoporosis medication and adherence, as well as reduced rates of fractures and hospitalization. Experience from different countries with various models of FLS has also proved its cost-effectiveness in care delivery. Hence, FLS should be adopted as a standard of care for patients who have sustained osteoporosis fractures in order to reduce the societal and economic cause of subsequent fractures.

Keywords Fracture liaison service · Osteoporosis · FLS

1 Introduction-Fragility Fracture and Fracture Liaison Service

Fragility fracture is defined as fracture that occurs following minimal trauma such as falling from standing height or less. These fractures typically indicate underlying osteoporosis or low bone density. Common fragility fractures include fracture distal radius, vertebra, proximal femur, proximal humerus and other sites. Individuals who experienced the first fracture are at increased risk of future fractures, especially within the first 2 years, hence the phrase "fracture begets fracture" [1]. However,

J.-K. Lee (✉)
Beacon Hospital, Petaling Jaya, Malaysia
e-mail: osteoporosis_jklee@yahoo.com

H.-M. Khor
Faculty of Medicine, University of Malaya, Kuala Lumpur, Malaysia

many patients do not receive treatment for secondary fracture prevention despite the recognition of this imminent risk [2]. To address this care gap, the implementation of appropriate secondary fracture prevention program had been carried out in many countries.

Fracture Liaison Service (FLS) is a specialized program that aims to prevent future fractures in patients with an initial fragility fracture. The FLS model involves identifying at-risk patients who have experienced a fragility fracture and then coordinating their care to improve bone health and prevent future fractures. One of the main functions of FLS is to ensure individuals who have suffered a fragility fracture receive appropriate evaluation and treatment for osteoporosis. It is a multidisciplinary care model targeted for fragility fracture care. The FLS team typically includes healthcare professionals such as nurses, physicians, and physical therapists who work together to identify and treat osteoporosis or other underlying risk factors that may be contributing to the patient's bone fragility.

The FLS program provides a comprehensive evaluation of the patient's lifestyle factors, medical history, and bone density that are influencing their bone health. The team will work with the patient to develop an individualized treatment plan that includes falls prevention, physical therapy, dietary changes, anti-osteoporosis medication and provide ongoing support to improve adherence to treatment and ensure appropriate monitoring and follow-up.

The ultimate goal of FLS is to improve patient outcomes and reduce the burden of fragility fractures on individuals and society as a whole. By identifying and treating patients at high risk of future fractures, FLS can help reduce patient's healthcare costs and improve their quality of life.

2 Key Components of Fracture Liaison Service in Optimizing Pharmacological Treatment in Osteoporosis

Fracture Liaison Service (FLS) plays a significant role in optimizing pharmacological treatment for osteoporosis. FLS can help facilitate the initiation of pharmacological treatment for osteoporosis by ensuring that patients receive timely evaluation and prescription for appropriate therapy. FLS can help monitor treatment response by facilitating regular BMD measurements and ensuring appropriate follow-up with healthcare providers. This can help ensure that patients are receiving optimal treatment and identify those who may require additional interventions. This comprehensive care is crucial as patients may fall through the osteoporosis treatment gap due to barriers such as underdiagnosis, lack of access to appropriate treatment and poor communication between healthcare providers [3, 4]. In order to address the osteoporosis treatment care gap, the core components of the Fracture Liaison Service should include the following:

- **Identification and assessment**: Patients who have experienced a fragility fracture are identified through the healthcare system and assessed to determine their risk of future fractures. The use of appropriate clinical risk prediction tool or referral for bone mineral density should be performed.
- **Treatment and management**: Patients who are found to be at risk of future fractures are provided with appropriate treatment and management, which include initiation of pharmacological treatment for osteoporosis, lifestyle changes and falls prevention intervention. FLS can facilitate clinicians in selecting the most appropriate pharmacological treatment for each patient based on individual characteristics such as age, sex, fracture history, comorbidities, and medication tolerance. Referral to other healthcare practitioners may be necessary to address health conditions that may contribute to future fractures.
- **Patient education and support**: Patients are provided with education and support to help them understand their condition and take an active role in their care. Education and counselling on the risk and benefit of different pharmacological intervention should be provided as this will also improve treatment adherence. This can include assistance with scheduling and medication reminders and follow-up monitoring to address any concerns or side effects with treatment.
- **Coordination of care**: The FLS team helps the patient to navigate the healthcare system to ensure that care assessment and treatment are communicated to the primary care physicians, who will monitor patient's response to treatment and continue long term follow-up.

3   Who are the Stake Holders and What are Their Roles in the FLS Program?

A Fracture Liaison Service (FLS) provides a multidisciplinary approach to the prevention and management of subsequent fragility fractures in individuals who have sustained an index fracture. Each member of the FLS team has a role in providing comprehensive and coordinated care to patients with fragility fractures. The implementation of FLS requires an FLS lead/champion (usually an orthopedic surgeon or physician) and a team of clinicians, allied health professions and administrators. An interdisciplinary collaboration among these stakeholders is essential where shared goals and responsibilities have a huge impact to patient outcomes (Table 1).

4   Models of Care in Fracture Liaison Service

There are several different models of fracture liaison service (FLS), which vary depending on factors such as healthcare system, patient population, and resource availability. The delivery of FLS include identification of fracture patients, organizes investigations, initiation of bone health intervention and follow-up plans [5].

Table 1 Examples of different members of the team contribute their expertise and skills in providing holistic care

Stakeholder	Roles and responsibilities
FLS coordinator	The coordinator is responsible for overseeing the day-to-day operations of the FLS, including case finding, risk assessment, and follow-up care. The coordinator is often the point of contact for patients and healthcare providers, and is responsible for ensuring that all members of the FLS are working together effectively
Orthopaedic surgeon	Orthopaedic surgeons are often the first point of contact for patients with fragility fractures, and play a crucial role in identifying and referring patients to the FLS. Early surgical intervention and providing the most stable fracture fixation, patients will be able to mobilize and ambulate earliest possible to prevent pressure sores, pneumonia and urinary infection secondary to immobilization in bed
Geriatrician	Geriatricians play the utmost important role in providing care for patients with frailty, cognitive impairment and older adults with multiple complex medical issues. Geriatric care in the inpatient setting includes medical optimisation, prevention of postoperative complications and implementation of falls intervention and osteoporosis management
Emergency medicine physician	Most individuals with fragility fractures present to the emergency department. Besides providing acute care in emergency department, emergency physicians play an important role in mobilizing the FLS team such as referral to FLS coordinator, orthopaedic surgeon and geriatrician to optimise patient's condition
Rheumatologist/ Endocrinologist	In many countries, endocrinologists and rheumatologists are FLS champions who lead the team and provide guidance on bone health assessment and pharmacological interventions
Primary care physician	Most individuals with fragility fractures identified by FLS will be communicated to their primary care physicians for continuation of care. Primary care physicians play an important role in follow up care as well as provide long term anti-osteoporosis management after patients are discharged from hospital
Rehabilitation medicine physicians and therapists	Successful surgery for fracture treatment requires tailored rehabilitation program to achieve optimal outcome. Rehabilitation medicine physicians, physical therapists and occupational therapists' involvement are crucial in assisting patients to return to their best functional status
Radiologist	Radiologists play an important role in supervising radiographers or technologists to perform high-quality bone mineral density (DXA) scan. A high-quality DXA helps to guide clinicians in making accurate risk assessment and clinical diagnosis. The identification of fragility fractures on radiological examinations assists in increasing detection rates of silent fractures which may be missed by clinicians

(continued)

Table 1 (continued)

Stakeholder	Roles and responsibilities
Pharmacist	Pharmacists conducts medication review, assess compliance, address lifestyle modification and recommend appropriate medication therapy to physicians as member of the FLS. Compliance and adherence to anti-osteoporosis medicines is very important to ensure the effectiveness of treatment
Dietician	Dietitians play an important role to determine adequate nutritional intake for older patients with poor bone health, osteoporosis and sarcopenia
Nursing staff	Nurses frequently take up the role as FLS coordinator who can provide patient education, coordinate care, and follow up visits with patients to ensure that they are receiving appropriate treatment and follow-up care
Administrator	The success of FLS is determined by ensuring that the team members have all the necessary resources, including funding, staffing, and equipment, to provide effective care. The administrators are responsible for regular monitoring and evaluation of the FLS standard of care
Patient and their families	Patients and their families play an important role in providing feedback on their experience of care under the FLS. They should hold a positive attitude and confidence towards the FLS care provided and adhere to the osteoporosis pharmacological treatment, falls prevention advise and appropriate dietary intake to improve musculoskeletal health

However, different FLS models of care may implement some or all of the components listed above depending on the specific setting and available resources. Here are some of the most common models of FLS in different settings:

- **Hospital-based FLS**: This model involves identifying and treating patients with fragility fractures while they are hospitalized, typically through a dedicated FLS team or a collaboration between orthopedic surgeons and other healthcare professionals.
- **Post-discharge FLS**: In this model, patients with fragility fractures are identified and treated after they have been discharged from the hospital, typically through primary care providers or community-based FLS teams.
- **Telehealth FLS**: This model involves using telehealth technologies such as video conferencing or remote monitoring to provide FLS services to patients who are unable to access in-person care. The use of telehealth or virtual fracture liaison clinics have increased since the COVID-19 pandemic.
- **Fracture prevention clinics**: These clinics specialize in identifying and treating patients at high risk of fragility fractures, typically through a multidisciplinary team of healthcare professionals who work together to provide comprehensive care.

- **Integrated FLS**: This model involves integrating FLS into existing healthcare systems, such as primary care practices or specialty clinics, to ensure that all patients with fragility fractures receive appropriate care.

The goal of FLS is to provide coordinated, comprehensive care to patients with fragility fractures in order to prevent future fractures and improve patient outcomes. The specific model of FLS may vary depending on local resources and patient needs.

5   The Role of FLS in Identifying Patients with High Fracture Risk and Treatment Options

An FLS can identify patients with high fracture risk through a comprehensive evaluation of their medical history, physical examination, and bone mineral density (BMD) testing. Patients who have already suffered a fragility fracture are at high risk of future fractures and are prime candidates for further evaluation and intervention. In addition, FLS can use risk assessment tools such as FRAX (Fracture Risk Assessment Tool) to identify patients who are at high risk of future fractures.

Once high-risk patients are identified, FLS can help in deciding on the appropriate treatment option for osteoporosis. This may include lifestyle modifications such as exercise and adequate calcium and vitamin D intake, as well as pharmacological interventions such as bisphosphonates, denosumab, romosozumab and teriparatide. FLS can also provide education and counseling on the benefits and risks of treatment options, as well as help patients navigate the healthcare system to ensure access to appropriate care.

By identifying patients at high risk of fractures and providing timely interventions, FLS can significantly reduce the burden of osteoporotic fractures and improve patient outcomes.

6   How FLS Helps Clinicians in Deciding Antiresorptive or Anabolic Agents for Treatment of Osteoporosis?

FLS can help clinicians in deciding between antiresorptive or anabolic agents for the treatment of osteoporosis by providing tailored assessments of fracture risk and other clinical factors, and by offering guidance on appropriate treatment options based on individual patient needs and preferences. FLS can also help patients navigate the healthcare system to ensure access to appropriate care and monitor treatment response to optimize outcomes.

Antiresorptive agents, such as bisphosphonates and denosumab, are the most commonly prescribed medications for osteoporosis. They work by slowing down the activity of bone-destroying cells called osteoclasts, thus reducing bone turnover and preserving bone mass. Antiresorptive agents are typically recommended for patients with osteoporosis who are at moderate to high risk of fractures.

Anabolic agents, such as teriparatide and abaloparatide, are a class of osteoporosis medications that stimulate the activity of bone-building cells called osteoblasts. Both teriparatide and abaloparatide have been shown to significantly increase bone mineral density and reduce fracture risk on vertebral and non-vertebral fractures. Both teriparatide and abaloparatide are administered for 24 months, and should be followed by anti-resorptive agents as the sequential therapy to maintain the bones gained. Romosozumab, a monoclonal antibody that binds to and inhibits sclerostin, appears to have dual actions by stimulating bone formation and reducing bone resorption. Romosozumab has been shown to reduce significantly new vertebral fractures, vertebral and non-vertebral fractures. Both anabolic agents and agent with dual actions are typically recommended for patients with severe osteoporosis with very high fracture risks or those who have failed to respond to other treatments.

7 What are the Guidelines Recommendation of Treatment Options Based on FLS?

7.1 European Society of Endocrinology

The Endocrine Society defines the following fracture risk categories [6]:

(a) Low risk includes no prior hip or spine fractures, a BMD T-score at the hip and spine both above -1.0, a 10-year hip fracture risk <3%, and 10-year risk of major osteoporotic fractures <20%;

(b) Moderate risk includes no prior hip or spine fractures, a BMD T-score at the hip and spine both above -2.5, and 10-year hip fracture risk <3% or risk of major osteoporotic fractures <20%;

(c) High risk includes a prior spine or hip fracture, or a BMD T-score at the hip or spine of -2.5 or below, or 10-year hip fracture risk $\geq$3%, or risk of major osteoporotic fracture risk $\geq$20%; and

(d) Very high risk includes multiple spine fractures and a BMD T-score at the hip or spine of -2.5 or below.

The Endocrine Society guidelines recommend that antiresorptive agents, such as bisphosphonates, denosumab, or selective estrogen receptor modulators (SERMs), should be used as first-line therapy for patients with high risk of fracture. Anabolic agents, such as teriparatide, romosozumab and abaloparatide should be reserved for patients with severe osteoporosis, very high risk of fragility fractures or those who have failed to respond to other treatments. The guidelines also recommend that treatment decisions should be individualized based on fracture risk, bone mineral density, comorbidities, and other clinical factors.

Based on the principles and rational of fracture liaison service (FLS), it is a comprehensive multidisciplinary approach of managing patients with pre-existing fragility fracture. These patients are categorized under the high risk or very high-risk categories for future fractures. FLS helps to guide treatment choices as it identifies patients with pre-existing fragility fracture, within the last one year with the highest imminent fracture risks. Pre-existing fragility fracture, single vertebral fracture, hip fracture or multiple vertebral fractures carry a different fracture risk which require anabolic agents as the first line of treatment.

FLS can be recommended for individuals aged 50 years or older who have had a fragility fracture, with evaluation of BMD and assessment of risk factors for future fractures. Monitoring treatment response with serial BMD measurements under the FLS program can help to facilitate adherence to therapy and follow-up. FLS should be implemented in healthcare systems to ensure that patients who have suffered fragility fractures are identified and receive appropriate evaluation and intervention for osteoporosis. The guidelines also recommend that postmenopausal women with osteoporosis should receive monitoring for treatment response, adjust treatment regimens as needed, and be provided with ongoing education and support.

7.2 American Association of Clinical Endocrinologist AACE

The American Association of Clinical Endocrinologists (AACE) also provides clinical practice guidelines for the management of osteoporosis [7].

The AACE guidelines categorize patients into high-risk category in those who presented with previous hip or spine fracture (>12 months), BMD T-score < −2.5 and FRAX >3% for hip and >20% Major Osteoporosis Fracture.

Patients are categorized as very high risk if there is a recent fracture (<12 months), multiple vertebral fractured, BMD T-score <−2.5 and any fracture(s), use of drugs that cause skeletal harm, BMD T-score <-3.0, FRAX score >4.5% (hip) or >30% (MOF) and high fall risk.

AACE guidelines recommend that antiresorptive agents, such as bisphosphonates, denosumab, or SERMs, should be used as first-line therapy for most patients in high-risk category. Anabolic agents, such as teriparatide, romosozumab and abaloparatide should be recommended as first line of treatment for patients with very high risk for future fractures or those who have failed to respond to other treatments.

The guidelines also emphasize the importance of individualizing treatment decisions based on a patient's fracture risk, BMD, comorbidities, and other clinical factors.

FLS can be utilized to identify individuals who have suffered fragility fractures and to perform appropriate evaluation for osteoporosis, including measurement of bone mineral density (BMD) and assessment of fracture risk. FLS can also be used to monitor treatment response, adjust treatment regimens as needed, and provide ongoing education and support to patients. FLS to help guide treatment choices and monitor treatment response over time.

7.3 International Osteoporosis Foundation (IOF) and European Society for Clinical and Economic Aspects of Osteoporosis and Osteoarthritis (ESCEO) Guideline

The International Osteoporosis Foundation (IOF) and European Society for Clinical and Economic Aspects of Osteoporosis and Osteoarthritis (ESCEO) have published clinical practice guidelines for the management of osteoporosis, which include recommendations for the use of FLS [8].

The IOF/ESCEO guidelines recommend that FLS should be implemented in healthcare systems to identify and manage patients at high risk for osteoporotic fractures, particularly those who have already suffered a fragility fracture. FLS should involve systematic identification of patients, evaluation of fracture risk, BMD testing, and appropriate initiation of pharmacological treatment and lifestyle interventions.

Regarding treatment options, the guidelines recommend that antiresorptive agents, such as bisphosphonates, denosumab, or SERMs, should be used as the first-line therapy for most patients with osteoporosis, and individuals with high fracture risk. Individuals categorized under the very high-risk category or those who have failed to respond to other treatments should receive anabolic agents, such as teriparatide, abaloparatide or romosozumab.

The guidelines emphasize the importance of individualized treatment decisions based on a patient's fracture risk, BMD, comorbidities, and other clinical factors, and the use of FLS to help guide treatment choices and monitor treatment response over time.

In summary, the IOF/ESCEO guidelines recommend the use of FLS to identify and manage patients with osteoporosis, including the use of antiresorptive agents as the first-line therapy for most patients and anabolic agents for selected patients based on individualized treatment decisions.

8 Clinical Effectiveness of FLS in Bridging the Osteoporosis Care Gap

There is a growing body of clinical evidence that supports the effectiveness of fracture liaison services (FLS) in improving the treatment of osteoporosis. Systematic review and meta-analysis by Li et al. in 16 studies showed that FLS programs reduced the risk of subsequent fractures by 30% with follow up period of at least 2 years duration. Most of the studies were comparing outcomes before and after implementation of FLS programs. In the subgroup analysis of pre-post FLS studies, there was reported reduction in mortality risk (OR: 0.65, 95% CI: 0.44–0.95, P = 0.03; heterogeneity: I2 = 95%) [9]. The findings were similar to findings from Nakayama et al., where there was a 40% reduction in re-fracture rates at the hip, spine, femur, pelvis or humerus at the FLS hospitals compared to non-FLS hospitals. The number needed to prevent one new fracture over 3 years was 20 [10].

In other systematic reviews, patients who received care from an FLS were more likely to be investigated with bone densitometry, initiated on pharmacological treatment for osteoporosis and taking medication within 6 months of their fracture compared to those who received usual care [11, 12]. The implementation of FLS has led to higher rates of medication adherence and persistence in patients with osteoporosis, compared to non-FLS controls [10].

Cost effectiveness of Fracture Liaison Service (FLS) programs have also been demonstrated in different countries, which include Canada, Australia, USA, UK, Europe, and Asia. In Japan, FLS is associated with an incremental cost-effectiveness ratio (ICER) ranging from $3023–$28,800 US dollars (USD) per quality-adjusted life year gained in comparison to usual care. The systematic review also found that FLS was cost-effective regardless of the intensity of the service or country where FLS was introduced [13].

9 Fracture Liaison Service in Improving Osteoporosis Treatment—Success Stories

Fracture Liaison Service (FLS) which was first described by McLellan and colleagues from Glasgow, Scotland have been implemented successfully in many parts of the world. Following that, many successful approaches had been adopted in different countries.

- **Glasgow, Scotland**: The Glasgow FLS program was established in 1999 and has since been recognized as a model for successful FLS implementation. This study evaluated the effectiveness of an FLS program implemented in Glasgow, Scotland, in reducing the risk of subsequent fractures in older patients with a wrist fracture. The program involved a multidisciplinary team approach and included patient education, medication management, and follow-up care. The study found that patients who received the FLS intervention had a significantly lower risk of subsequent fractures compared to those who received usual care. The program was also associated with improved rates of osteoporosis treatment initiation. The Glasgow FLS program has since been recognized as a model for successful FLS implementation [14].
- **Kaiser Permanente, United States**: Kaiser Permanente is a large healthcare system in the United States that has implemented a successful FLS program across multiple locations. The program is based on a multidisciplinary team approach involving orthopedic surgeons, endocrinologists, primary care physicians, and other specialists. The program has been associated with significant reductions in subsequent fractures and improved rates of osteoporosis treatment initiation, as well as cost savings due to reduced hospitalizations [15, 16].
- **Canada**: Majumdar SR et al. reported on a cost-effectiveness analysis of an osteoporosis case manager program for patients with hip fractures in Canada. The program included patient education and counseling, medication review and

adjustment, and follow-up BMD testing. It was found that the program was cost-effective and led to improved rates of osteoporosis medication use and BMD testing, as well as reduced rates of re-fracture and hospitalizations [17].

- **Singapore**: The Osteoporosis Patient Targeted and Integrated Management for Active Living (OPTIMAL) is a Singapore Ministry of Health-funded secondary fracture prevention program. The Osteoporosis and Bone Metabolism Unit at the National University Hospital in Singapore, one of the participating hospitals in the OPTIMAL program established an FLS program in 2008. The program is based on a multidisciplinary team approach involving different departments such as endocrinology, emergency medicine, orthopedics, rheumatology, geriatrics, obstetrics and gynecology, internal medicine, family medicine, and physical medicine and rehabilitation. The program has been associated with improved rates of osteoporosis treatment initiation with a 72% compliance rate of up to two years among patients enrolled in the program. The authors suggest that the FLS program is an effective way to close the osteoporosis care gap in Singapore [18].
- **Taiwan**: The first Fracture Liaison Service program in Taiwan was established in National Taiwan University Hospital. The model of service delivery is accordance to the consensus recommendations of a task force from the Taiwan Osteoporosis Association on the implementation of an FLS program for osteoporosis in Taiwan. The recommendations include a multidisciplinary approach to patient care, risk assessment and treatment algorithms, patient education and monitoring, and data collection and analysis. Subsequently this model of care has become the primary framework for many other fracture prevention programs throughout the country. Since then, the number of FLS programs has increased to 22 sites in 2018 as the highest FLS coverage countries in the Asia Pacific region [19].
- **South Korea**: The Korean Society for Bone and Mineral Research established the Korea Fracture Liaison Service (KFLS) Committee in 2018 which aimed to provide recommendations for the implementation of FLS and suggest future directions for effective treatment of osteoporosis and fragility fractures and prevention of secondary fractures [20, 21].

These successful FLS programs demonstrate the effectiveness of a multidisciplinary team approach, patient education and support, and coordinated care for patients with osteoporosis and subsequent fractures. They serve as models for FLS implementation in other Asia Pacific countries and emphasize the importance of collaboration between healthcare professionals, patients, and healthcare systems in reducing the burden of osteoporosis and subsequent fractures in the region.

10 Setting Up FLS

Setting up an FLS program requires a multifaceted approach that involves demonstrating evidence, emphasizing benefits, building partnerships, addressing concerns, and involving patients. By engaging stakeholders and demonstrating the effectiveness of FLS programs, you can build support and enthusiasm for this important initiative.

Convincing stakeholders to participate in your Fracture Liaison Service (FLS) program can be challenging, but here are some strategies that may be helpful:

- **Share evidence and benefits**: Share clinical evidence and research studies that demonstrate the benefits of FLS programs in reducing subsequent fractures, improving patient outcomes, and reducing healthcare costs. This can help to persuade stakeholders of the importance and effectiveness of FLS programs.
- **Foster partnerships and address concerns**: Work to build partnerships and collaborations with stakeholders, including healthcare providers, patients, and community organizations.

  Listen to stakeholders' concerns and address them in a constructive manner. For example, if a stakeholder is concerned about the cost of implementing an FLS program, provide evidence of cost savings associated with reduced subsequent fractures.
- **Pilot program**: Consider piloting the FLS program in a smaller setting or with a specific patient population to demonstrate its effectiveness and build support among stakeholders.
- **Involve patients**: Involve patients in the development and implementation of the FLS program to help build support and engagement. This can also help to ensure that the program is patient-centered and meets the needs of the patient population.

11 Fracture Liaison Service Best Practice Framework

With varying models of FLS delivered globally, the implementation of certain standards are necessary to ensure care is consistent with best practice in order to achieve the effective outcome. The presence of a FLS Framework at the national, regional or global level will allow all existing or potential FLS to follow a common guideline to standardize the approach in terms of planning, execution, work flow, data collection and analysis of all FLS at different levels. This will allow comparisons of quality and standard of FLS in the same region with a standardized method. As a minimal common dataset will be included in all different frameworks, all FLS will be working on the same platform where data collection and analysis are standardized.

There are several examples of Fracture Liaison Service (FLS) frameworks available:

(a) **The International Osteoporosis Foundation (IOF) Fracture Liaison Service (FLS) Framework**: This is a resource developed by the International Osteoporosis Foundation (IOF) launched in 2012 to guide healthcare professionals and organizations in the implementation and optimization of FLS programs. The framework provides a comprehensive set of guidelines and recommendations for the development, implementation, and evaluation of FLS programs, including key components such as case finding, assessment and management of fracture risk, patient education and support, and quality improvement measures. A total of 13 best practice standards were identified which function as performance indicators for a FLS. The same standards were reviewed by clinical experts from the Asia Pacific region who concluded that they were generally applicable and required only a few advanced clarifications to support quality improvement of FLS in the Asia Pacific region [22, 23].

(b) **Fracture Liaison Service Implementation (FLS-I) Toolkit**: The FLS-I Toolkit is a resource developed by the Royal Osteoporosis Society (ROS) in the UK to support the implementation of FLS programs. The toolkit provides guidance on how to set up and run an FLS program, including a step-by-step guide, case studies, and resources for patient education and engagement. The clinical standards of FLS is measured against the 5iQ model developed by the ROS and patient level performance indicators by the Royal College of Physicians Fracture Liaison Service Database (FLSDB) audit, a mandatory audit for all secondary care hospitals in England and Wales [24].

Fracture Liaison Service Framework is a highly useful and effective healthcare model that can improve patient outcomes, reduce healthcare costs, and enhance patient satisfaction. It is increasingly being adopted as a standard of care for patients with osteoporosis, and is likely to continue to play a key role in the management and prevention of fractures in at-risk populations.

12 How to Perform KPI for My FLS?

The key performance indicators (KPIs) for an FLS program can vary depending on the specific goals of the program, but here are some examples of common metrics that may be measured:

- **Fracture capture rate**: This measures the percentage of eligible patients who are identified and enrolled in the FLS program. This KPI can help assess the effectiveness of the program in identifying patients with fragility fractures.
- **Osteoporosis treatment initiation rate**: This measures the percentage of patients who are started on osteoporosis treatment after being identified through the FLS program. This KPI can help assess the effectiveness of the program in ensuring that patients receive appropriate treatment for osteoporosis.
- **Follow-up rate**: This measures the percentage of patients who attend scheduled follow-up appointments after being enrolled in the FLS program. This KPI can help assess the program's effectiveness in engaging and retaining patients in care.

- **Secondary fracture rate**: This measures the rate of subsequent fractures among patients enrolled in the FLS program. This KPI can help assess the effectiveness of the program in preventing subsequent fractures and improving patient outcomes.
- **Cost savings**: This measures the cost savings associated with the FLS program, including reduced hospitalizations and other healthcare costs. This KPI can help assess the cost-effectiveness of the program and its impact on the healthcare system.

13 What Are the Possible Pitfalls of Fracture Liaison Service?

While Fracture Liaison Service (FLS) programs have been shown to be effective in improving treatment rate, treatment compliance and adherence, reducing subsequent fractures and improving patient outcomes, there are also potential pitfalls that should be considered, including:

- **Limited resources**: Implementing and maintaining an FLS program can be resource-intensive, and require significant investment in terms of staff time, equipment, and infrastructure. This can be a challenge for healthcare systems that are already stretched thin.
- **Implementation barriers**: Implementing an FLS program can face barriers such as inadequate staffing, resistance to change, limited funding, and insufficient data management systems. Addressing these barriers may require significant effort and resources.
- **Cost**: FLS programs require resources, including staff time, equipment, and medication costs. It is important to consider the cost-effectiveness of the program and ensure that it is financially sustainable.
- **Workload**: FLS programs can place an additional workload on healthcare providers, particularly in the initial stages of implementation. It is important to ensure that staff are adequately trained and resourced to manage the workload.
- **Patient participation**: Some patients may not wish to participate in FLS programs or may not be able to comply with the requirements of the program, such as attending regular appointments or adhering to medication regimes.
- **Patient adherence**: FLS programs require patient engagement and adherence to be effective. Patients may face barriers to attending appointments, taking medications, and making lifestyle changes, which can reduce the effectiveness of the program.
- **Limited access to care**: Some patients may have limited access to healthcare services due to factors such as geographic location, transportation barriers, and financial constraints. This can make it difficult for them to participate in an FLS program.

- **Cultural and social barriers**: Cultural and social barriers, such as language and communication barriers, may also limit the effectiveness of FLS programs in some populations.
- **Integration with existing healthcare systems**: FLS programs may require changes to existing healthcare systems, including electronic medical records and referral processes. It is important to ensure that the program is integrated smoothly into existing systems to avoid disruption to patient care.
- **Sustainability**: An FLS program requires long-term commitment and sustainability. Without ongoing support and resources, the program may not be able to achieve the desired outcomes.

It is important to address these potential pitfalls and develop strategies to overcome them in order to achieve the best outcomes for patients and healthcare systems.

14 Conclusion

Fracture Liaison Service (FLS) programs are comprehensive care models that aim to improve the identification, evaluation, and management of osteoporosis and subsequent fractures. Evidence suggests that FLS programs are effective interventions for improving treatment rate, treatment compliance and adherence, identifying the right groups of patients receiving the appropriate types of treatments, reducing rates of subsequent fractures, improving patient outcomes, and are cost-effective.

The implementation of FLS programs requires a multidisciplinary team approach, involving healthcare professionals such as orthopedic surgeons, primary care physicians, endocrinologists, and other specialists. Patient education and support are also key components of FLS programs, which can help improve adherence to osteoporosis medications and lifestyle modifications.

The implementation of FLS programs is an important step towards reducing the burden of osteoporosis and subsequent fractures, and improving the quality of life for patients. With increasing adoption of FLS programs as a standard of care, healthcare systems can provide more comprehensive care to patients with osteoporosis and reduce the societal and economic impact of subsequent fractures.

References

1. Kanis JA, Johnell O, De Laet C et al (2004) A meta-analysis of previous fracture and subsequent fracture risk. Bone 35:375–382
2. Kung AW, Fan T, Xu L et al (2013) Factors influencing diagnosis and treatment of osteoporosis after a fragility fracture among postmenopausal women in Asian countries: a retrospective study. BMC Womens Health 13:7
3. Elliot-Gibson V, Bogoch ER, Jamal SA et al (2004) Practice patterns in the diagnosis and treatment of osteoporosis after a fragility fracture: a systematic review. Osteoporos Int 15:767–778

4. Giangregorio L, Papaioannou A, Cranney A et al (2006) Fragility fractures and the osteoporosis care gap: an international phenomenon. Semin Arthritis Rheum 35:293–305
5. Ganda K, Puech M, Chen JS et al (2013) Models of care for the secondary prevention of osteoporotic fractures: a systematic review and meta-analysis. Osteoporos Int 24:393–406
6. Shoback D, Rosen CJ, Black DM et al (2020) pharmacological management of osteoporosis in postmenopausal women: an endocrine society guideline update. J Clin Endocrinol Metab 105:dgaa048.
7. Camacho PM, Petak SM, Binkley N et al (2020) American association of clinical endocrinologists/American college of endocrinology clinical practice guidelines for the diagnosis and treatment of postmenopausal osteoporosis-2020 update. Endocr Pract 26:1
8. Kanis JA, Cooper C, Rizzoli R et al (2019) European guidance for the diagnosis and management of osteoporosis in postmenopausal women. Osteoporos Int 30:3–44
9. Li N, Hiligsmann M, Boonen A et al (2021) The impact of fracture liaison services on subsequent fractures and mortality: a systematic literature review and meta-analysis. Osteoporos Int 32:1517–1530
10. Nakayama A, Major G, Holliday E et al (2016) Evidence of effectiveness of a fracture liaison service to reduce the re-fracture rate. Osteoporos Int 27:873–879
11. Wu CH, Tu ST, Chang YF et al (2018) Fracture liaison services improve outcomes of patients with osteoporosis-related fractures: a systematic literature review and meta-analysis. Bone 111:92–100
12. Sale JE, Beaton D, Posen J et al (2011) Systematic review on interventions to improve osteoporosis investigation and treatment in fragility fracture patients. Osteoporos Int 22:2067–2082
13. Wu CH, Kao IJ, Hung WC et al (2018) Economic impact and cost-effectiveness of fracture liaison services: a systematic review of the literature. Osteoporos Int 29:1227–1242
14. McLellan AR, Gallacher SJ, Fraser M et al (2003) The fracture liaison service: success of a program for the evaluation and management of patients with osteoporotic fracture. Osteoporos Int 14:1028–1034
15. Dell R (2011) Fracture prevention in Kaiser Permanente Southern California. Osteoporos Int 22:457–460
16. Dell R, Greene D, Schelkun SR et al (2008) Osteoporosis disease management: the role of the orthopaedic surgeon. J Bone Joint Surg Am 90:188–194
17. Majumdar SR, Lier DA, Beaupre LA et al (2009) Osteoporosis case manager for patients with hip fractures: results of a cost-effectiveness analysis conducted alongside a randomized trial. Arch Intern Med 169:25–31
18. Chandran M, Tan MZ, Cheen M et al (2013) Secondary prevention of osteoporotic fractures–an "OPTIMAL" model of care from Singapore. Osteoporos Int 24:2809–2817
19. Chang LY, Tsai KS, Peng JK et al (2018) The development of Taiwan fracture liaison service network. Osteoporos Sarcopenia 4:47–52
20. Cha YH, Ha YC, Lim JY (2019) Establishment of fracture liaison service in Korea: where is it stand and where is it going? J Bone Metab 26:207–211
21. Lim JY, Kim YY, Kim JW et al (2023) Fracture liaison service in Korea: 2022 position statement of the Korean society for bone and mineral research. J Bone Metab 30:31–36
22. Akesson K, Marsh D, Mitchell PJ et al (2013) Capture the fracture: a best practice framework and global campaign to break the fragility fracture cycle. Osteoporos Int 24:2135–2152
23. Chan DD, Chang LY, Akesson KE et al (2018) Consensus on best practice standards for fracture liaison service in the Asia-Pacific region. Arch Osteoporos 13:59
24. Royal Osteoporosis Society (2022) Fracture liaison service (FLS) toolkit. https://theros.org.uk/healthcare-professionals/fracture-liaison-services/implementation-toolkit/. Accessed 28 Apr 2023

Bone Considerations in Hip and Knee Arthroplasty

Cass Nakasone and Sian Yik Lim

Abstract In this chapter, we discuss the topic of bone health in total joint arthroplasty patients. Osteoporosis and osteopenia is common in patients undergoing total joint arthroplasty. We discuss several bone health considerations, research topics that are of interest to the orthopedic surgeon. We also discuss optimization of bone health, as well as studies of the utility of osteoporosis medications in total joint arthroplasty patients.

Keywords Total joint replacement · Osteoporosis · Osteopenia · Periprosthetic fractures

1 Introduction

Osteoporosis is a common condition affecting almost 10 million individuals 50 years and older in the United States [1]. Both osteoporosis and osteoarthritis increase in prevalence as individuals age, with severe osteoarthritis patients being treated with knee or hip arthroplasty. Knee and hip arthroplasties have increased in recent years. While bone quality affects osteointegration and durability of the prosthesis [2], metabolic bone diseases in total joint arthroplasty patients are undertreated and under-screened [3].

C. Nakasone · S. Y. Lim (✉)
Hawaii Pacific Health Medical Group, Honolulu, HI, USA
e-mail: limsianyik@gmail.com

S. Y. Lim
Bone and Joint Center, Pali Momi Medical Center, 98-1079 Moanalua Road, Suite 300, Aiea, HI 96701, USA

2 Osteoporosis and Osteopenia in Total Joint Replacement Patients

Osteoporosis and osteopenia are common in patients undergoing total joint replacement [4]. Based on recent studies, approximately 60–70% of patients undergoing TJA have osteopenia or osteoporosis [4]. In a study involving 53 patients undergoing cementless total hip replacement, Makinen et al. found that 28% had osteoporosis while 45% had osteopenia [5]. In a meta-analysis of total joint replacement patients, the pooled prevalence of osteoporosis was 24.8%, while osteopenia was 38.5% [2]. The high prevalence of osteoporosis and osteopenia, highlights the importance of addressing bone health early as patients age in particular in patients with arthritis where joint arthroplasty may be anticipated. It is important to note that osteoporosis medications require time to work, and optimization of bone health is a process that spans over several years.

3 Orthopedic Surgeon Considerations: Bone Quality, Prosthesis Design, Cement Use, and Surgical Techniques

The bone quality of a patient before arthroplasty has implications for orthopedic surgeons. In general, bone integration into the porous implant is important for good surgical outcomes. In both hip and knee arthroplasties, cemented implants have historically been considered ideal particularly in patients with poor bone quality [4]. However, use of cementless implants have continually increased with high early success rates demonstrated even with short proximally porous coated press-fit tapered stems in total hip arthroplasties [5–7]. While cemented implants are still considered the gold standard for patients with poor bone quality, it remains unclear if this remains true. A recent study reported higher rates of aseptic loosening with high-viscosity cement use in total knee replacements [8]. Furthermore, removal of fully cemented implants can be technically challenging in osteopenic bone and can represent a technical disadvantage if a revision procedure is required [5].

Prosthesis design—use of cement and surgical techniques—has demonstrated variable effects on adjacent bone and it remains unclear whether cemented or uncemented techniques are best. Kamath et al. compared the periprosthetic bone mineral density of patients who underwent cemented (30 patients) or uncemented (30 patients) total knee arthroplasty and found decreased bone mineral density in adjacent bone of both groups [9]. Four years following surgery, no significant differences in periprosthetic bone mineral density could be found [9]. While the authors allude that more costly uncemented implants did not show a clear advantage, statistical significance favoring uncemented femoral components was approached calling into question whether a larger sample size would have demonstrated an advantage of uncemented femoral implants [9]. Resulting changes in mechanical alignment following arthroplasty often alter mechanical loads on surrounding bone and can result in focal areas of periprosthetic bone loss. This adaptive biologic process, known as "stress shielding" causes bone to resorb in areas of abnormally low physiologic loads resulting from specific implant designs, changes in mechanical alignment or more likely, a combination of both [10]. Newer generations of uncemented femoral implants have porous coatings limited to the proximal regions of the stem to reduce the problem of proximal femoral bone loss. Limiting bony fixation to the proximal femur replicates normal physiologic femoral loading thus reducing bone loss as a result of stress shielding with such designs showing excellent early results [6, 7]. Periprosthetic bone loss can also occur due to osteolysis. Osteolysis can occur as a result of macrophage phagocytosis of particulate wear particles (produced in all mechanical bearings) which activate osteoclast-mediated bone resorption [11].

4   Periprosthethic Fractures

Periprosthetic fractures are catastrophic failures usually requiring operative solutions and occur in bone adjacent to orthopedic implants (replacement or internal fixation device) [12]. Osteoporosis is a significant associated risk factor for periprosthetic fracture [13, 14]. Patients who suffer periprosthetic fractures are usually frail and often have concurrent osteopenia or osteoporosis [4, 12]. The periprosthetic fracture incidence after primary total hip replacement is estimated to be approximately 0.9% and 0.6% following primary total knee replacement [14]. Periprosthetic fracture incidence is significantly greater after revision hip arthroplasty (4.2%) and following revision knee arthroplasty (1.7%) [14]. Unfortunately, periprosthetic fractures are challenging to treat and are associated with significantly greater morbidity and mortality as well as prolonged recovery periods [14].

5   Optimization of Bone Health/Osteoporosis Medications in Total Joint Arthroplasty Patients

Due to the negative impact of osteoporosis on knee or hip arthroplasty success, initiation of bone health optimization should be considered far in advance of any knee or hip arthroplasty procedure for optimal outcomes when possible. While many questions remain, few studies address the utility of osteoporosis treatment following joint arthroplasty. By inhibiting osteoclasts, bisphosphonates possibly facilitate early bony ingrowth, may decrease periprosthetic bone loss and lower rates of revision surgeries [11]. Several bisphosphonates (risedronate, alendronate and zoledronic acid) have been reported to increase periprosthetic bone mineral density and/or decrease periprosthetic bone loss following hip and knee replacement [15–17]. Bisphosphonates have also been associated with improved outcomes and lower rates of all-cause revisions in patients being treated [18, 19]. The benefits of bisphosphonates must be weighed against the risk of jaw osteonecrosis, atypical fractures, and possibly a slight increase in periprosthetic fractures, especially in younger patients [11]. Some authors have recommended limiting treatment with bisphosphonates to 1 year regardless of bone density at the time of surgery [11].

Several newer osteoporosis medications have also demonstrated success in improving bone quality following knee or hip arthroplasty. Denosumab administration was recently reported to result in greater femoral periprosthetic bone mineral density after cementless total hip arthroplasty [20] and appeared to prevent periprosthetic bone mineral density loss in the tibial metaphysis following total knee arthroplasty [21]. In a proof-of-concept study, a single injection of denosumab reportedly reduced osteoclast activity at the osteolysis membrane-bone interface by 83% compared to patients who received placebo [22]. Osteoanabolic agents offer an alternative to bisphosphonates and have demonstrated clinical promise. In a randomized controlled trial, teriparatide was shown to be as effective as alendronate in preventing periprosthetic bone loss after total hip replacement [23].

6 Summary

In summary, the problem of osteopenia and osteoporosis at the time of hip and knee arthroplasty remains poorly addressed. Despite the associated risks, surgeons will be required to perform hip and knee arthroplasties in patients presenting with significant osteopenia. The choice of implant, best method of fixation and surgical approach remain unclear and surgeons will need to make decisions with incomplete data using their best clinical judgement. Method and timing of bone health optimization is an evolving concept, requiring significant future research. A multispecialty approach involving the primary care physician, orthopedic surgeon and most importantly, a metabolic bone specialist may help optimize outcomes in patients undergoing hip or knee arthroplasty.

References

1. Cosman F, de Beur SJ, LeBoff MS et al (2014) Clinician's guide to prevention and treatment of osteoporosis. Osteoporos Int 25:2359–2381
2. Xiao PL, Hsu CJ, Ma YG et al (2022) Prevalence and treatment rate of osteoporosis in patients undergoing total knee and hip arthroplasty: a systematic review and meta-analysis. Arch Osteoporos 17:16
3. Wang Z, Levin JE, Amen TB et al (2022) Total joint arthroplasty and osteoporosis: looking beyond the joint to bone health. J Arthroplasty 37:1719-1725.e1
4. Blankstein M, Lentine B, Nelms NJ (2020) The use of cement in hip arthroplasty: a contemporary perspective. J Am Acad Orthop Surg 28:e586–e594
5. Russell LA (2013) Osteoporosis and orthopedic surgery: effect of bone health on total joint arthroplasty outcome. Curr Rheumatol Rep 15:371

6. Nishioka ST, Andrews SN, Mathews K et al (2022) Varus malalignment of short femoral stem not associated with post-hip arthroplasty fracture. Arch Orthop Trauma Surg 142:3533–3538

7. Andrews S, Harbison GJ, Hasegawa I et al (2020) Perioperative fracture risk and two-year survivorship of a short tapered femoral stem following direct anterior approach cementless total hip arthroplasty with a fracture table. J Hip Surg 04:033–037. https://doi.org/10.1055/s-0040-170853

8. Buller LT, Rao V, Chiu YF et al (2020) Primary total knee arthroplasty performed using high-viscosity cement is associated with higher odds of revision for aseptic loosening. J Arthroplasty 35:S182–S189

9. Kamath S, Chang W, Shaari E et al (2008) Comparison of peri-prosthetic bone density in cemented and uncemented total knee arthroplasty. Acta Orthop Belg 74:354–359

10. Millis DL (2014) Responses of musculoskeletal tissues to disuse and remobilization. In: Millis D, Levine D (eds) Canine rehabilitation and physical therapy, 2nd edn. W.B. Saunders, Pennsylvania, pp 92–153

11. McDonald CL, Lemme NJ, Testa EJ et al (2022) Bisphosphonates in total joint arthroplasty: a review of their use and complications. Arthroplast Today 14:133–139

12. Marsland D, Mears SC (2012) A review of periprosthetic femoral fractures associated with total hip arthroplasty. Geriatr Orthop Surg Rehabil 3:107–120

13. Caruso G, Milani L, Marko T et al (2018) Surgical treatment of periprosthetic femoral fractures: a retrospective study with functional and radiological outcomes from 2010 to 2016. Eur J Orthop Surg Traumatol 28:931–938

14. Bottle A, Griffiths R, White S et al (2020) Periprosthetic fractures: the next fragility fracture epidemic? A national observational study. BMJ Open 10:e042371

15. Su J, Wei Y, Li XM et al (2018) Efficacy of risedronate in improving bone mineral density in patients undergoing total hip arthroplasty: a meta-analysis of randomized controlled trials. Medicine 97:e13346

16. Soininvaara TA, Jurvelin JS, Miettinen HJ et al (2002) Effect of alendronate on periprosthetic bone loss after total knee arthroplasty: a one-year, randomized, controlled trial of 19 patients. Calcif Tissue Int 71:472–477

17. Gao J, Gao C, Li H et al (2017) Effect of zoledronic acid on reducing femoral bone mineral density loss following total hip arthroplasty: a meta-analysis from randomized controlled trails. Int J Surg 47:116–126

18. Namba RS, Inacio MC, Cheetham TC et al (2016) Lower total knee arthroplasty revision risk associated with bisphosphonate use, even in patients with normal bone density. J Arthroplasty 31:537–541

19. Ro DH, Jin H, Park JY et al (2019) The use of bisphosphonates after joint arthroplasty is associated with lower implant revision rate. Knee Surg Sports Traumatol Arthrosc 27:2082–2089

20. Nyström A, Kiritopoulos D, Ullmark G et al (2020) denosumab prevents early periprosthetic bone loss after uncemented total hip arthroplasty: results from a randomized placebo-controlled clinical trial. J Bone Miner Res 35:239–247

21. Murahashi Y, Teramoto A, Jimbo S et al (2020) Denosumab prevents periprosthetic bone mineral density loss in the tibial metaphysis in total knee arthroplasty. Knee 27:580–586

22. Mahatma MM, Jayasuriya RL, Hughes D et al (2021) Effect of denosumab on osteolytic lesion activity after total hip arthroplasty: a single-centre, randomised, double-blind, placebo-controlled, proof of concept trial. Lancet Rheumatol 3:e195–e203

23. Kobayashi N, Inaba Y, Uchiyama M et al (2016) Teriparatide versus alendronate for the preservation of bone mineral density after total hip arthroplasty: a randomized controlled trial. J Arthroplasty 31:333–338

Utility of Osteoporosis Medications in Palliative Care and Oncology

Liang Yik Lim, Chin Heng Fong, Sui Keat Tan, Cheen Leng Lee, Ying Ying Sum, and Jun Sian Lim

Abstract Osteoporosis medications are used for a range of purposes in Oncology and Palliative care. These include their use in the prevention of skeletal-related events, adjuvant cancer treatment, malignant hypercalcemia, and cancer pain. In this chapter, we aim to provide the reader with background knowledge of its use for these indications. We provide information on dose, frequency, and duration of osteoporosis medications, which differ when used for indications other than osteoporosis. We also discuss the use of medications for osteoporosis in cancer patients, and the deprescribing osteoporosis medications in the poor prognosis setting. We hope this chapter will broaden the readers' understanding of the use of osteoporosis medications in Oncology and Palliative care.

1 Introduction

Osteoporosis medications slow bone loss by affecting bone metabolism. Over the past decades, its effect on bone metabolism has been shown to lead to beneficial effects in cancer treatment. Osteoporosis medications have been shown to prevent skeletal related events in patients with bone metastasis and multiple myeloma, and improve cancer treatment outcomes, especially in breast cancer. Osteoporosis medications also play an essential role in the treatment of malignant hypercalcemia. Their use for indications other than osteoporosis often lead to different dosages, frequency,

L. Y. Lim (✉)
Palliative Care Unit, Penang General Hospital, George Town, Malaysia

C. H. Fong · C. L. Lee · Y. Y. Sum
Department of Oncology and Palliative Care, Penang General Hospital, George Town, Malaysia

S. K. Tan
Hematology Unit, Penang General Hospital, George Town, Malaysia

J. S. Lim
Department of Internal Medicine, Penang General Hospital, George Town, Malaysia

A. H. Choi and S. Yik Lim (eds.), *Pharmacological Interventions for Osteoporosis*, Tissue Repair and Reconstruction, https://doi.org/10.1007/978-981-99-5826-9_8

and duration. In addition, cancer and its treatment often increase the risk for osteoporosis and fractures, and the management of osteoporosis in cancer requires special consideration.

Osteoporosis medications use can also lead to the palliation of bone pain, in conjunction with analgesics. In the poor prognosis setting, decisions need to be made regarding the continuation of osteoporosis medications, regardless of whether it is used for osteoporosis or other indications, and this requires skillful consideration of the evolving balance between risk and benefit. This chapter aims to discuss important aspects when using osteoporosis medications in Oncology and Palliative care.

2 Bone-Targeted Agents in the Prevention of Skeletal Related Events (SRE) in Cancer

2.1 Background

The skeleton is the most common organ to be affected by metastatic cancer, especially for malignancies arising from breast, prostate, thyroid, lung and kidneys [1].

According to a retrospective study, bone was the most common site of metastatic disease with 69% of patients dying with breast cancer having radiological evidence of skeletal metastasis before death. Bone was also the most common site of first distant relapse, representing almost half of all first relapses. The median survival in patients with first relapse in bone was 20 months. However, patients with bone only metastasis survive longer, with a median survival of 24 months. Among these patients, 29% developed one or more complications from bone metastasis, namely hypercalcemia, pathological fracture (16%) or spinal cord compression (3%) [2].

In contrast, 85–100% of patients who die of prostate carcinoma have bone metastases. Patients with more advanced local disease and higher grade of cancer have greater risk of developing bone mets [3]. The median time for initial skeletal event after diagnosis of bone metastasis was 9.5 months [4]. The median survival for all patients with bone metastases who receive hormonal therapy is 30–35 months [3]. More than 30% of these patients will develop skeletal complications including vertebral deformity or collapse requiring spinal instrumentation. Weight bearing bones had a higher risk of pathological fractures. Patients with bone pain at baseline had higher risk of eventual skeletal complications [4].

Besides bone metastases, skeletal complications can also occur as a treatment related adverse events. Patients with prostate cancer who were treated with long term hormonal deprivation therapy were more likely to have osteoporotic fractures than malignant fractures and overall had a 12.2-fold relative risk of bone fracture compared with age-matched men without prostate carcinoma. Likewise, bone mineral density differed significantly between patients treated with hormonal ablation compared with normal age-matched controls and this disparity increased with the length of treatment [3].

2.2 Efficacy Data

Since the 1980s, small studies of first-generation bisphosphonates in breast and prostate cancers with bone metastases have shown benefits in terms of bone pain and progression [5, 6]. Subsequent large clinical trials found that oral clodronate significantly reduced the incidence of vertebral fractures and skeletal morbidity associated with breast cancer [7].

By the 1990s, clinical studies on second generation bisphosphonates showed that intravenous pamidronate reduced skeletal morbidity rate, incidence and time to first skeletal complication significantly compared to placebo [8]. Long term outcome studies confirmed the benefits of pamidronate supplement to antineoplastic therapies in preserving bone integrity and for palliation [8], leading to its approval for the treatment of complications arising from bone metastases in breast cancer [9].

The advent of the more potent third generation bisphosphonate led to clinical trials comparing zoledronate with pamidronate in breast cancer, and with placebo in other solid tumors. Zoledronate was the first and only bisphosphonate to be proven effective in patients with all types of bone lesions, from osteolytic to osteoblastic. Zoledronate was found to be superior to pamidronate in breast cancer, and superior to placebo in prostate, lung and other cancers with regards to long term safety and efficacy in managing skeletal morbidities [9–14].

Denosumab is the latest bone-targeted agent that acts by binding to RANKL, decreasing osteoclastic formation and activity, demonstrating rapid suppression of bone turnover in patients with bone metastasis. Compared with zoledronate in patients with solid tumors, denosumab exhibited superior potency, significantly delaying time to skeletal-related events, and pain improvement, while maintaining similar safety profiles [9, 15].

2.3 Clinical Applications

The current ESMO guideline recommends for bone targeting agents to be initiated at diagnosis of bone metastasis and be considered throughout the course of the disease, especially for breast and prostate cancers, whether patients are symptomatic or not. In other cancers, namely lung, renal and other solid tumors, zoledronate or denosumab is also recommended if there are clinically significant bone metastases and patients are expected to survive for more than 3 months [16].

A meta-analysis on dosing frequency compared 4-weekly and 12-weekly intravenous bisphosphonates and found that 12-weekly was non-inferior to 4-weekly administration, with no significant difference in the incidence of skeletal-related events, and drug related complications. Nonetheless, there appears to be possible increase in serious SRE requiring bone surgery in the 12-weekly schedule, suggesting that patients with multiple bone metastasis should initiate with the 4-weekly schedule for 3 to 6 months before converting to the extended treatment schedule [16, 17].

The following antiresorptive treatments are recommended in cancer patients for bone metastases:

Bone-targeted agents	Indication	Typical administration
Denosumab	All solid tumours	120 mg s.c. every 4 weeks
Zoledronate	All solid tumours	4 mg i.v. every 3–4 weeks
Pamidronate	Breast cancer	90 mg i.v. every 3–4 weeks
Clodronate	Osteolytic lesions	1600 mg p.o./day
Ibandronate	Breast cancer	50 mg p.o./day
		6 mg i.v./month
Bone health in cancer: ESMO Clinical Practice Guidelines supplementary material [16]		

Patients with multiple bone metastases should be treated with a bone-targeted agent according to a standard dosing schedule and consider converting to an extended schedule if patients achieve complete response or good partial response. On the other hand, patients with oligometastatic bone disease can be considered for 12-weekly zoledronate on initiation, with the possibility of interrupting bone-targeted agents after 2 years of treatment if patients achieve complete response or good partial response and are deemed to have low risk of bone complications. Patients should resume bone-targeted agents upon disease progression [16].

3 Osteoporosis Medications as Adjuvant Treatment in Cancer

In addition to prevention and treatment of SREs, osteoporosis medications have a role in improving survival outcomes in patients with breast cancer, when used in the adjuvant setting with no clinical evidence of bone metastasis. The bone microenvironment is known to be a site for cancer cell survival and proliferation, providing a theoretical basis for treatment.

Over the past two decades, clinical trials have evaluated the benefits of bisphosphonates as an adjuvant treatment for breast cancer [18]. The Early Breast Cancer Clinical Trials Group (EBCCTG) conducted a meta-analysis of 18,766 women randomized in trials to either receive adjuvant bisphosphonates in addition to the standard breast cancer therapy or standard breast cancer therapy alone and found significant reduction in breast cancer recurrence rates in the bone, but had a minimal impact on other breast cancer outcomes. However, further subgroup analysis showed that in postmenopausal women, whether natural menopause or induced menopause by suppression of ovarian function, benefit was observed in breast cancer mortality, bone recurrence and all-cause mortality with bisphosphonate therapy at 10 years, with an absolute risk reduction of 3.3%, 2.2% and 2.3%, respectively [19]. Hence, bisphosphonate therapy may be more beneficial as an adjuvant therapy in post-menopausal women. This advantage was demonstrated regardless of the type or schedule of bisphosphonate use.

Different bisphosphonate regimens, such as oral clodronate, oral ibandronate and intravenous zoledronate, have been studied as adjuvant therapy in breast cancer treatment with similar efficacy [20]. SWOG 0307 trial randomized women with early breast cancer to receive either oral clodronate, oral ibandronate, or intravenous zoledronate for three years (Table 1). There were no significant differences in the 5-year disease-free survival or overall survival, independent of age or tumor subtypes [21, 22].

The optimal dose and duration of bisphosphonates as adjuvant therapy in women with early breast cancer have not been clearly defined [23]. The Cancer Care Ontario and the American Society of Clinical Oncology (CCO/ASCO) guideline recommends intravenous zoledronate or oral clodronate for postmenopausal women with breast cancer who are eligible for adjuvant bisphosphonate therapy, at doses shown in Table 2 [24]. To date, there is lack of evidence to support more frequent dosing with monthly intravenous zoledronate and daily oral ibandronate. The phase III SUCCESS A trial found no benefit with extending therapy from two to five years [25].

Apart from bisphosphonates, use of denosumab in early breast cancer offers promising prospects. However, the phase III ABSCG-18 and the D-CARE adjuvant trials yielded conflicting results. Because of this, the use of denosumab in the adjuvant setting for breast cancer requires further investigation [26].

Bisphosphonates have a dual function in the context of early breast cancer: inhibition of metastasis and prevention of treatment-induced bone loss [16]. Recommendations for its use in the adjuvant setting are summarized in Fig. 1.

Table 1 Type and schedule of bisphosphonate agents in SWOG 0307 trial

Bisphosphonate agents	Route and dose	Schedule
Clodronate	PO 1600 mg	Daily for 3 years
Ibandronate	PO 50 mg	Daily for 3 years
Zoledronate	IV 4 mg	Monthly for 6 months, then every 3 months for 2.5 years

Table 2 Recommended dosage and schedule of bisphosphonate agents as adjuvant therapy in early breast cancer

Bisphosphonate agents	Route and dose	Schedule
Zoledronate	IV 4 mg	Every 6 months for 3–5 years
Clodronate	PO 1600 mg	Daily for 2–3 years

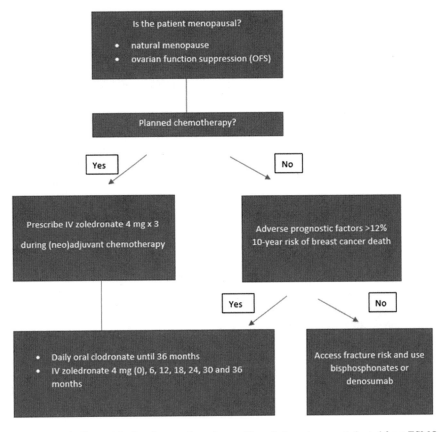

Fig. 1 Use of adjuvant bisphosphonates in patients with early breast cancer (adapted from ESMO Clinical Practice Guidelines)

4   Usage of Osteoporosis Agents in Multiple Myeloma

Multiple Myeloma is a plasma cell dyscrasia associated with high risk of skeletal related events including pathological fractures and spinal cord compression [27, 28]. The pathophysiology of multiple myeloma-related bone disease involves multiple myeloma cells interaction with bone cells including osteocytes, osteoblasts, and osteoclasts [29]. Multiple myeloma-induced osteocyte apoptosis leads to a favorable niche for myeloma cell homing. Osteocytes also produce soluble factors, including receptor activator of NF-κB, sclerostin, and Dickkopf-1 which promote osteoclast activity and impair osteoblast maturation [30]. Increased osteoclast activity is driven by the activation of the RANK–RANKL signaling system [31].

For many years, bisphosphonates have been the gold standard for multiple myeloma-related bone disease prevention and treatment [32]. However, an enhanced understanding of the underlying pathophysiology of multiple myeloma-related bone disease has led to clinical investigations of other targeted agents such as denosumab in the treatment of multiple myeloma. Bisphosphonates are pyrophosphate analogues that bind to exposed areas of hydroxyapatite crystals during the bone remodeling process. Osteoclasts endocytose bisphosphonates, which are potent inhibitors of the intracellular farnesyl pyrophosphate synthase, lead to osteoclast apoptosis and prevention of bone loss [33]. Bisphosphonates are indicated in all patients with active multiple myeloma, regardless of the presence or absence of multiple myeloma related bone disease on imaging studies. Zoledronic acid 4 mg administered intravenously every 3–4 weeks over 15 min infusion, and pamidronic acid 30 mg or 90 mg administered every 3–4 weeks over 45 min (for 30 mg) or 2 h (for 90 mg) are recommended for skeletal-related event prevention. Dose adjustments for bisphosphonates are essential in case of renal impairment.

Denosumab is a fully human and highly specific monoclonal IgG2 antibody against RANKL [28]. Denosumab initiates the physiological effect of osteoprotegerin (also known as TNFRSF11B) by inhibiting RANKL interaction with RANK, ultimately decreasing bone resorption. Denosumab is recommended both for the treatment of newly diagnosed multiple myeloma and for patients with relapsed or refractory multiple myeloma with evidence of multiple myeloma-related bone disease. Denosumab is equivalent to zoledronic acid in delaying the time to the first skeletal-related event after multiple myeloma diagnosis. Denosumab may be preferable over zoledronic acid in patients with renal dysfunction. Denosumab can be considered for patients who have creatinine clearance lower than 30 mL/min under close monitoring. Denosumab can be also administered in patients with hypercalcemia related to myeloma, especially in patients who are refractory to zoledronic acid administration. Denosumab is administered as a subcutaneous injection of 120 mg at monthly intervals.

Bisphosphonate or denosumab is the standard of care for the treatment of multiple myeloma-related bone disease [34]. The decision to choose one bone-targeted agent over another should include consideration of multiple factors such as cost, convenience, patient preference and toxicity profile. Zoledronic acid should be the preferred treatment option for patients who do not have imaging findings for multiple myeloma-related bone disease, whereas denosumab should be the preferred treatment option for patients with renal impairment [34].

5 Osteoporosis Management in Cancer Patients

Cancer and its treatment can have a profound impact on bone health, especially in older patients, as bone fragility increases with age. However, preventive efforts among cancer patients have not been adequately described [35], especially in premenopausal women and men aged <50 [36].

Some treatments used in oncology disease can cause hypogonadism, such as aromatase inhibitors, chemotherapy, Gonadotrophin Releasing Hormone (GnRH) Analogues/antagonists, glucocorticoids, surgical castration and pelvic radiotherapy.

- Age
- Female
- Body mass index
- Previous fragility fracture
- Parental history of hip fracture
- Alcohol intake
- Thoracic kyphosis
- Height loss (>4 cm)
- Falls and Frailty
- Inflammatory disease: *e.g.*, ankylosing spondylitis, other inflammatory arthritis, connective tissue diseases, systemic lupus erythematosus
- Endocrine disease: *e.g.*, type I and II diabetes mellitus, hyperparathyroidism, hyperthyroidism, hypogonadism, Cushing's disease/syndrome
- Hematological disorders/malignancy *e.g.*, multiple myeloma, thalassemia
- Muscle disease: *e.g.*, myositis, myopathies and dystrophies, sarcopenia
- Lung disease: *e.g.*, asthma, cystic fibrosis, chronic obstructive pulmonary disease
- HIV
- Neurological/ psychiatric disease *e.g.*, Parkinson's disease and associated syndromes, multiple sclerosis, epilepsy, stroke, depression, dementia
- Nutritional deficiencies: calcium, vitamin D
- Bariatric surgery and other conditions associated with intestinal malabsorption.
- Medications, *e.g.*:
  - Some immunosuppressants (calmodulin/calcineurin phosphatase inhibitors)
  - Hyperthyroidism hormone treatment (levothyroxine and/or liothyronine)
  - Drugs affecting gonadal hormone production (aromatase inhibitors, androgen deprivation therapy, medroxyprogesterone acetate, gonadotrophin hormone-releasing agonists, gonadotrophin hormone receptor antagonists)
  - Some diabetes drugs (*e.g.*, thiazolidinediones)
  - Some antiepileptics (*e.g.*, phenytoin and carbamazepine)

Fig. 2 Clinical risk factors for osteoporosis (adopted from UK clinical guideline for the prevention and treatment of osteoporosis [16])

This contributes to increase in bone-remodeling, progressive reduction of bone mineral density (BMD) and increased fracture risk [37]. Cancer patients may also have pre-existing risk factors for osteoporosis and fractures [36] (Fig. 2).

ESMO guidelines on bone health in cancer recommends all oncology patients receiving treatments that are known to adversely affect bone health ensure a sufficient dietary calcium/vitamin D intake and supplement these as necessary (Vitamin D 1000–2000 IU daily and Calcium 500–1000 mg daily), moderation of alcohol consumption ($\leq$2 units daily), smoking cessation, regular weight bearing and muscle strengthening exercise [16]. Cancer patients receiving oncological treatments that affect bone health and have additional risk factors for osteoporosis should undergo timely fracture risk assessment, using Fracture Risk Assessment Tool (FRAX) and bone mineral density testing with a DXA scan. If identified, osteoporosis should be promptly treated with an antiresorptive agent (Table 3) [16, 36]. Figure 3 provides a management algorithm for managing bone health during cancer treatment [16].

Table 3 Agents for osteoporosis in cancer

Bone-targeted agents	Route and dose	Schedule	Precaution
Bisphosphonates			
Alendronate	PO 70 mg	Weekly	Common side effects are upper gastrointestinal and bowel disturbance. Recommended taken after an overnight fast and swallow in upright position Risedronate—contraindicated in GFR ≤30 mL/min Alendronate and Ibandronate—cautioned use in GFR ≤30 mL/min
Risedronate	PO 35 mg	Weekly	
Ibandronate	PO 150 mg	Monthly	
Zoledronate	IV 5 mg	Yearly	Contraindicated in GFR ≤35 mL/min
Monoclonal antibody			
Denosumab	SC 60 mg	6 monthly	

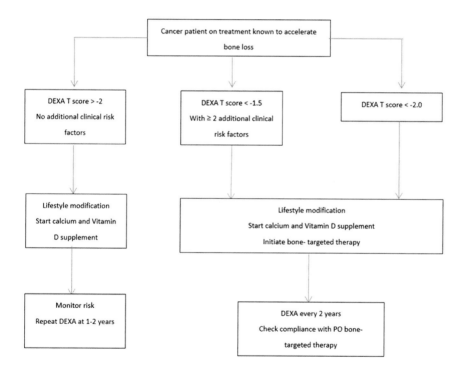

Fig. 3 Adopted algorithm from ESMO for managing bone health during cancer treatment

6 Use of Osteoporosis Agents in Malignant Hypercalcemia

Hypercalcemia is a common complication of patients with solid tumors and hematological malignancies. The most common mechanism is accelerated bone resorption driven by direct osteolysis (bone metastasis), cytokine-driven osteoclasts maturation, or secretion of parathyroid hormone-related protein (PTH-rP) by tumors. Occasionally, other clinical factors such as calcium and vitamin D supplementation or immobilization may simultaneously contribute to the development of hypercalcemia. Prognosis of hypercalcemia depends on its cause. Hypercalcemia due to malignancy is a marker of poor prognosis, especially when options for further effective anti-tumor therapy are limited [38–40].

Hypovolemia and renal impairment are common complications of severe hypercalcemia, and these may in turn impair calcium clearance through the kidneys. Hence, isotonic saline is often the first step in the treatment of severe hypercalcemia. Osteoporosis agents, with their effects on reducing osteoclasts activity and calcium metabolism, are often used in conjunction with hydration in the treatment of hypercalcemia. They are usually utilized in severe hypercalcemia, as indicated by markedly raised calcium level >3.5 mmol/L (>14 mg/dL) or altered sensorium (lethargy, stupor), which is more common if there is a rapid rise in calcium levels. ECG changes of hypercalcemia is also an indication for treatment.

Various bisphosphonates have been shown to be effective in hypercalcemia. They include: IV zoledronic acid [39], IV pamidronate [41–43], IV Ibandronate [40]. Zoledronic acid has been found to be superior to pamidronate, with higher response rates [39]. Normalization of calcium levels usually occurs after 2–10 days of bisphosphonate use. The duration of serum calcium control usually ranges from 18 to 43 days and is also longer with zoledronic acid compared to pamidronate [39]. Renal toxicity of bisphosphonates has been noted [44]. This is an important consideration in patients with pre-existing renal disease or markedly raised serum creatinine. Bisphosphonates may still be used in patients with mild acute renal impairment due to hypercalcemia, after risk benefit considerations and adequate hydration. In renal impairment, doses may need to be reduced and infusion duration prolonged to minimize risk. In addition to renal toxicity, hypocalcaemia is more common when bisphosphonate is used in renal impairment. Table 4 provides information on dose and infusion time.

Table 4 Doses of Osteoporosis Agents in Malignant Hypercalcemia

	Dose, infusion time	Time to repeat dose	Renal impairment
Zoledronic acid	4 mg over 15 min	7 days if hypercalcemia persists	Use not recommended unless benefit outweigh risk. Some experts administer 2 to 4 mg over 30–60 min
Pamidronate	60–90 mg over 2 to 24 h	7 days	Clinical judgement if benefit outweigh risk. 30–45 mg over 4 h
Calcitonin	4–8 units/kg every 12 h for 24–48 h	–	–
Denosumab	SC 60 mg Refractory hypercalcemia: SC 120 mg	7 days Monthly in long term therapy	Not cleared by the kidneys. No dose adjustments.

Antiresorptive RANK ligands inhibitors, such as denosumab, also have been used in the treatment of hypercalcemia of malignancy. It is usually used when there are contraindications to bisphosphonate use, such as allergy or severe renal impairment. Denosumab is not cleared by the kidneys. Denosumab is also used when hypercalcemia is refractory to bisphosphonates. In this setting, response rates of 64% have been reported [45]. When used, denosumab typically lowers calcium within few days to 2 weeks after dosing.

Calcitonin, a naturally occurring hormone in the body utilized to treat osteoporosis, is used in the early hours to days of treatment when severe life-threatening hypercalcemia necessitates rapid reduction of calcium levels [46]. Calcitonin has a rapid onset of action (within 4–6 h). However, effects of calcitonin are modest (lowering calcium levels by 0.3–0.5 mmol/l) and tachyphylaxis occurs within days [47]. IM or SC injections is the typical route of administration when treating hypercalcemia. The nasal route is not used due to its slow action. Calcitonin is often used in combination with bisphosphonates, which provide a more sustained control over calcium levels.

Information on the administration of osteoporosis agents for malignant hypercalcemia is given in Table 4.

Treatment of hypercalcemia related to malignancies should involve cancer targeted treatments whenever possible, usually after the resolution of hypercalcemia. Given malignant hypercalcemia is a marker of poor prognosis, its occurrence should trigger an assessment of palliative care needs, with additional measures to address these needs.

7 Deprescribing Osteoporosis Agents in Poor Prognosis Settings

Among people with limited prognosis (<1 year), deprescribing of osteoporosis agents used for fracture prevention should be considered. Goals of treatment often change when prognosis is poor, moving away from preventive treatment to a focus on the quality of life and reducing treatment burden. Osteoporosis medication administration may be more burdensome for patients with declining health and function. Requirements such as the need for patients to be seated upright to ingest medication, with a cup of water, or making trips to a health care facility for subcutaneous or intravenous injections, can be demanding for a patient with ill health. There is often also uncertainty about benefit and risk from continued treatment, as most clinical trials did not include patients with ill health.

Bisphosphonates treatment can be withdrawn without an increased risk of fractures during the 12–18 months after withdrawal, especially when treatment has been given for a few years. Similarly, stopping calcium and vitamin D supplementation did not increase non-vertebral fractures in the first 2 years after withdrawal [48]. However, discontinuation of denosumab is associated with rebound effect, which may lead to increased risk of fractures. Osteoporosis agents, when used for indications such as hypercalcemia, pain and fracture prevention from malignant bone metastasis are often continued late into the disease trajectory, due to beneficial effects on symptom control and prevention of impactful clinical complications [49].

Evaluation for deprescribing requires an understanding and discussion of the larger picture of a person's health status and prognosis. Shared decision making often involves understanding a patient's goals and values and if invited, making a medical recommendation that is aligned to the patient's goals and values, and eliciting their thoughts and feelings about the recommendation. While many patients welcome a discussion about reducing medication load, some patients may become concerned about "being given up on" or stopping a treatment that was meant to be "lifelong" [50]. Discussing deprescribing often involves opening discussion about the patient's general health status. The emotional context of the conversation should be attended to.

One of the barriers to deprescribing is shared care among many clinicians. Deprescribing when the prognosis is poor is often best initiated by the clinician who has a broad understanding of the physical, psychological, social and spiritual state of the patient, in collaboration with the primary prescriber. When there are concerns about contradicting other clinicians, communication among healthcare professionals is essential in facilitating the process.

8 Treatment of Pain Using Anti-osteoporosis Agents

Osteoporosis agents are used for the treatment of pain in several bone conditions, such as bone malignant disease, osteoporosis, osteoarthritis, and Paget disease.

8.1 Malignant Diseases in the Bone

Malignant diseases in the bone often result in significant pain. Osteoclasts inhibitors, such as bisphosphonates and denosumab have been shown to reduce pain in malignant bone diseases, particularly in metastatic breast cancer, prostate cancer, and multiple myeloma. Analgesic effects were observed when bisphosphonates were administered across months to years for skeletal-related-events prevention [51, 52]. One small study investigated pamidronate use in patients with poorly-controlled osteolytic bone pain and found significant analgesic effects after one week [53]. Effects of bisphosphonates and denosumab on pain could be small and varied, depending on disease type and osteolytic/osteoblastic lesions [54]. Analgesics, such as opioids and NSAIDs, and radiotherapy remain the mainstay of therapy in malignant bone disease, with osteoclastic inhibitors a useful adjunct.

8.2 Osteoporosis

Osteoporosis causes pain by two mechanisms, either from fragility fractures or osteo-porotic pathology without evidence of fracture (known as osteoporotic pain) [55]. Osteoporotic pain is often multifactorial, and include undetectable microfractures, altered bone structure or the osteoporotic state itself. Osteoporosis agents reduce the risk of painful fractures. Studies also demonstrate that they can mitigate pain in patients with chronic osteoporotic pain. A few Japanese studies demonstrate that alendronate is effective in mitigating osteoporotic pain and improving quality of life of postmenopausal women with osteoporotic pain [56–58]. Limited evidence also suggest that Denosumab reduces pain in post-menopausal women with osteoporosis [59]. Other osteoporosis agents, such as vitamin D, Teriparatide, strontium, calcitonin, and raloxifene are also reported to decrease chronic pain in patients with osteoporosis.

8.3 Paget Disease

Paget disease is a disease of bone remodelling with a predilection for the skull, thoracolumbar spine, pelvis and long bones. It results in deformity, fractures, arthritis, and pain. It is thought to be a disease of osteoclasts. Bisphosphonates are effective in bone pain and improve quality of life in patients with Paget disease [60].

8.4 Osteoarthritis

Several studies show that bisphosphonates reduce pain and improve function in osteoarthritis [61, 62]. Inhibition of osteoclasts could slow down subchondral bone remodelling. At this point, their use in osteoarthritis remains investigational.

References

1. Coleman RE (1997) Skeletal complications of malignancy. Cancer 80:1588–1594
2. Coleman RE, Rubens RD (1987) The clinical course of bone metastases from breast cancer. Br J Cancer 55:61–66
3. Carlin BI, Andriole GL (2000) The natural history, skeletal complications, and management of bone metastases in patients with prostate carcinoma. Cancer 88:2989–2994
4. Berruti A, Dogliotti L, Bitossi R et al (2000) Incidence of skeletal complications in patients with bone metastatic prostate cancer and hormone refractory disease: predictive role of bone resorption and formation markers evaluated at baseline. J Urol 164:1248–1253
5. Elomaa I, Blomqvist C, Porkka L et al (1987) Treatment of skeletal disease in breast cancer: a controlled clodronate trial. Bone 8 Suppl 1:S53–S56

6. Elomaa I, Kylmälä T, Tammela T et al (1992) Effect of oral clodronate on bone pain. A controlled study in patients with metastic prostatic cancer. Int Urol Nephrol 24:159–166

7. Paterson AH, Powles TJ, Kanis JA et al (1993) Double-blind controlled trial of oral clodronate in patients with bone metastases from breast cancer. J Clin Oncol 11:59–65

8. Theriault RL, Lipton A, Hortobagyi GN et al (1999) Pamidronate reduces skeletal morbidity in women with advanced breast cancer and lytic bone lesions: a randomized, placebo-controlled trial. Protocol 18 Aredia Breast Cancer Study Group. J Clin Oncol 17:846–854

9. von Moos R, Costa L, Gonzalez-Suarez E et al (2019) Management of bone health in solid tumours: from bisphosphonates to a monoclonal antibody. Cancer Treat Rev 76:57–67

10. Rosen LS, Gordon D, Kaminski M et al (2001) Zoledronic acid versus pamidronate in the treatment of skeletal metastases in patients with breast cancer or osteolytic lesions of multiple myeloma: a phase III, double-blind, comparative trial. Cancer J 7:377–387

11. Rosen LS, Gordon D, Tchekmedyian NS et al (2004) Long-term efficacy and safety of zoledronic acid in the treatment of skeletal metastases in patients with nonsmall cell lung carcinoma and other solid tumors: a randomized, Phase III, double-blind, placebo-controlled trial. Cancer 100:2613–2621

12. Rosen LS, Gordon D, Kaminski M et al (2003) Long-term efficacy and safety of zoledronic acid compared with pamidronate disodium in the treatment of skeletal complications in patients with advanced multiple myeloma or breast carcinoma: a randomized, double-blind, multicenter, comparative trial. Cancer 98:1735–1744

13. Saad F, Gleason DM, Murray R et al (2002) A randomized, placebo-controlled trial of zoledronic acid in patients with hormone-refractory metastatic prostate carcinoma. J Natl Cancer Inst 94:1458–1468

14. Saad F, Gleason DM, Murray R et al (2004) Long-term efficacy of zoledronic acid for the prevention of skeletal complications in patients with metastatic hormone-refractory prostate cancer. J Natl Cancer Inst 96:879–882

15. Sun L, Yu S (2013) Efficacy and safety of denosumab versus zoledronic acid in patients with bone metastases: a systematic review and meta-analysis. Am J Clin Oncol 36:399–403

16. Coleman R, Hadji P, Body JJ et al (2020) Bone health in cancer: ESMO Clinical Practice Guidelines. Ann Oncol 31:1650–1663

17. Yang M, Yu X (2020) Management of bone metastasis with intravenous bisphosphonates in breast cancer: a systematic review and meta-analysis of dosing frequency. Support Care Cancer 28:2533–2540

18. Strobl S, Wimmer K, Exner R et al (2018) Adjuvant bisphosphonate therapy in postmenopausal breast cancer. Curr Treat Options Oncol 19:18. https://doi.org/10.1007/s11864-018-0535-z

19. Early Breast Cancer Trialists' Collaborative Group (EBCTCG) (2015) Adjuvant bisphosphonate treatment in early breast cancer: meta-analyses of individual patient data from randomised trials. Lancet 386:1353–1361

20. Brufsky A, Mathew A (2020) Bisphosphonate choice as adjuvant therapy for breast cancer: does it matter? J Natl Cancer Inst 112:659–660

21. Gralow JR, Barlow WE, Paterson AHG et al (2020) Phase III randomized trial of bisphosphonates as adjuvant therapy in breast cancer: S0307. J Natl Cancer Inst 112:698–707

22. Ottewell P, Wilson C (2019) Bone-targeted agents in breast cancer: do we now have all the answers? Breast Cancer 13:1178223419843501. https://doi.org/10.1177/1178223419843501

23. Friedl TWP, Fehm T, Müller V et al (2021) Prognosis of patients with early breast cancer receiving 5 years vs 2 years of adjuvant bisphosphonate treatment: a phase 3 randomized clinical trial. JAMA Oncol 7:1149–1157

24. Dhesy-Thind S, Fletcher GG, Blanchette PS et al (2017) Use of adjuvant bisphosphonates and other bone-modifying agents in breast cancer: a cancer care Ontario and American society of clinical oncology clinical practice guideline. J Clin Oncol 35:2062–2081

25. Lu M, Ren B, Rao L (2022) Optimal duration of adjuvant bisphosphonate treatment for high-risk early breast cancer: Results from a SUCCESS trial. Thorac Cancer 13:519–520

26. Brufsky A, Mathew A (2020) Adjuvant bisphosphonate therapy in early-stage breast cancer-Treating the soil to kill the seed. Breast J 26:65–68

27. Terpos E, Berenson J, Cook RJ et al (2010) Prognostic variables for survival and skeletal complications in patients with multiple myeloma osteolytic bone disease. Leukemia 24:1043–1049
28. Terpos E, Ntanasis-Stathopoulos I, Dimopoulos MA (2019) Myeloma bone disease: from biology findings to treatment approaches. Blood 133:1534–1539
29. Terpos E, Ntanasis-Stathopoulos I, Gavriatopoulou M et al (2018) Pathogenesis of bone disease in multiple myeloma: from bench to bedside. Blood Cancer J 8:7. https://doi.org/10.1038/s41408-017-0037-4
30. McDonald MM, Reagan MR, Youlten SE et al (2017) Inhibiting the osteocyte-specific protein sclerostin increases bone mass and fracture resistance in multiple myeloma. Blood 129:3452–3464
31. Qiang YW, Chen Y, Stephens O et al (2008) Myeloma-derived Dickkopf-1 disrupts Wnt-regulated osteoprotegerin and RANKL production by osteoblasts: a potential mechanism underlying osteolytic bone lesions in multiple myeloma. Blood 112:196–207
32. Terpos E, Morgan G, Dimopoulos MA et al (2013) International Myeloma Working Group recommendations for the treatment of multiple myeloma-related bone disease. J Clin Oncol 31:2347–2357
33. van Beek E, Pieterman E, Cohen L et al (1999) Farnesyl pyrophosphate synthase is the molecular target of nitrogen-containing bisphosphonates. Biochem Biophys Res Commun 264:108–111
34. Terpos E, Zamagni E, Lentzsch S et al (2021) Treatment of multiple myeloma-related bone disease: recommendations from the Bone Working Group of the International Myeloma Working Group. Lancet Oncol 22:e119–e130
35. Fujihara N, Fujihara Y, Hamada S et al (2021) Current practice patterns of osteoporosis treatment in cancer patients and effects of therapeutic interventions in a tertiary center. PLoS ONE 16:e0248188. https://doi.org/10.1371/journal.pone.0248188
36. Gregson CL, Armstrong DJ, Bowden J et al (2022) UK clinical guideline for the prevention and treatment of osteoporosis. Arch Osteoporos 17:58. https://doi.org/10.1007/s11657-022-01061-5
37. Ottanelli S (2015) Prevention and treatment of bone fragility in cancer patient. Clin Cases Miner Bone Metab 12:116–129
38. Ramos REO, Perez Mak M, Alves MFS et al (2017) Malignancy-related hypercalcemia in advanced solid tumors: survival outcomes. J Glob Oncol 3:728–733
39. Major P, Lortholary A, Hon J et al (2001) Zoledronic acid is superior to pamidronate in the treatment of hypercalcemia of malignancy: a pooled analysis of two randomized, controlled clinical trials. J Clin Oncol 19:558–567
40. Pecherstorfer M, Steinhauer EU, Rizzoli R et al (2003) Efficacy and safety of ibandronate in the treatment of hypercalcemia of malignancy: a randomized multicentric comparison to pamidronate. Support Care Cancer 11:539–547
41. Gucalp R, Ritch P, Wiernik PH et al (1992) Comparative study of pamidronate disodium and etidronate disodium in the treatment of cancer-related hypercalcemia. J Clin Oncol 1:134–142
42. Nussbaum SR, Younger J, Vandepol CJ et al (1993) Single-dose intravenous therapy with pamidronate for the treatment of hypercalcemia of malignancy: comparison of 30-, 60-, and 90-mg dosages. Am J Med 95:297–304
43. Wimalawansa SJ (1994) Optimal frequency of administration of pamidronate in patients with hypercalcaemia of malignancy. Clin Endocrinol 41:591–595
44. Palmer S, Tillman F 3rd, Sharma P et al (2021) Safety of intravenous bisphosphonates for the treatment of hypercalcemia in patients with preexisting renal dysfunction. Ann Pharmacother 55:303–310
45. Hu MI, Glezerman IG, Leboulleux S et al (2014) Denosumab for treatment of hypercalcemia of malignancy. J Clin Endocrinol Metab 99:3144–3152
46. Walker MD, Shane E (2022) Hypercalcemia: a review. JAMA 328:1624–1636
47. Chevallier B, Peyron R, Basuyau JP et al (1988) Calcitonine humaine dans les hypercalcémies néoplasiques. Résultats d'un essai prospectif randomisé [Human calcitonin in neoplastic hypercalcemia. Results of a prospective randomized trial]. Presse Med 17:2375–2377

48. Brokaar EJ, van den Bos F, Visser LE et al (2022) Deprescribing in older adults with cancer and limited life expectancy: an integrative review. Am J Hosp Palliat Care 39:86–100
49. Hedman C, Frisk G, Björkhem-Bergman L (2022) Deprescribing in Palliative Cancer Care. Life 12:613. https://doi.org/10.3390/life12050613
50. Tjia J, Kutner JS, Ritchie CS et al (2017) Perceptions of statin discontinuation among patients with life-limiting illness. J Palliat Med 20:1098–1103
51. Stopeck AT, Lipton A, Body JJ et al (2010) Denosumab compared with zoledronic acid for the treatment of bone metastases in patients with advanced breast cancer: a randomized, double-blind study. J Clin Oncol 28:5132–5139
52. Mhaskar R, Kumar A, Miladinovic B et al (2017) Bisphosphonates in multiple myeloma: an updated network meta-analysis. Cochrane Database Syst Rev 12:CD003188. https://doi.org/10.1002/14651858.CD003188.pub4
53. Petcu EB, Schug SA, Smith H (2002) Clinical evaluation of onset of analgesia using intravenous pamidronate in metastatic bone pain. J Pain Symptom Manage 24:281–284
54. Hendriks LE, Hermans BC, van den Beuken-van Everdingen MH et al (2016) Effect of bisphosphonates, denosumab, and radioisotopes on bone pain and quality of life in patients with non-small cell lung cancer and bone metastases: a systematic Review. J Thorac Oncol 11:155–173
55. Orita S, Inage K, Suzuki M et al (2017) Pathomechanisms and management of osteoporotic pain with no traumatic evidence. Spine Surg Relat Res 1:121–128
56. Dursun N, Dursun E, Yalçin S (2001) Comparison of alendronate, calcitonin and calcium treatments in postmenopausal osteoporosis. Int J Clin Pract 55:505–509
57. Kawate H, Ohnaka K, Adachi M et al (2010) Alendronate improves QOL of postmenopausal women with osteoporosis. Clin Interv Aging 5:123–131
58. Iwamoto J, Makita K, Sato Y et al (2011) Alendronate is more effective than elcatonin in improving pain and quality of life in postmenopausal women with osteoporosis. Osteoporos Int 22:2735–2742
59. Petranova T, Sheytanov I, Monov S et al (2014) Denosumab improves bone mineral density and microarchitecture and reduces bone pain in women with osteoporosis with and without glucocorticoid treatment. Biotechnol Biotechnol Equip 28:1127–1137
60. Reid IR, Miller P, Lyles K et al (2005) Comparison of a single infusion of zoledronic acid with risedronate for Paget's disease. N Engl J Med 353:898–908
61. Spector TD, Conaghan PG, Buckland-Wright JC eet al (2005) Effect of risedronate on joint structure and symptoms of knee osteoarthritis: results of the BRISK randomized, controlled trial [ISRCTN01928173]. Arthritis Res Ther 7:R625–R633
62. Laslett LL, Doré DA, Quinn SJ et al (2012) Zoledronic acid reduces knee pain and bone marrow lesions over 1 year: a randomised controlled trial. Ann Rheum Dis 71:1322–1328

Printed in the United States
by Baker & Taylor Publisher Services